Scarecrow Professional Intelligence Education Series

Series Editor: Jan Goldman

In this post–September 11, 2001, era there has been rapid growth in the number of professional intelligence training and educational programs across the United States and abroad. Colleges and universities, as well as high schools, are developing programs and courses in homeland security, intelligence analysis, and law enforcement, in support of national security.

The Scarecrow Professional Intelligence Education Series (SPIES) was first designed for individuals studying for careers in intelligence and to help improve the skills of those already in the profession; however, it was also developed to educate the public in how intelligence work is conducted and should be conducted in this important and vital profession.

1. *Communicating with Intelligence: Writing and Briefing in the Intelligence and National Security Communities*, by James S. Major. 2008.
2. *A Spy's Résumé: Confessions of a Maverick Intelligence Professional and Misadventure Capitalist*, by Marc Anthony Viola. 2008.
3. *An Introduction to Intelligence Research and Analysis*, by Jerome Clauser, revised and edited by Jan Goldman. 2008.
4. *Writing Classified and Unclassified Papers for National Security: A Scarecrow Professional Intelligence Educational Series Manual*, by James S. Major. 2009.
5. *Strategic Intelligence: A Handbook for Practitioners, Managers, and Users*, revised edition by Don McDowell. 2009.
6. *Partly Cloudy: Ethics in War, Espionage, Covert Action, and Interrogation*, by David L. Perry. 2009.
7. *Tokyo Rose / An American Patriot: A Dual Biography*, by Frederick P. Close. 2010.
8. *Ethics of Spying: A Reader for the Intelligence Professional*, edited by Jan Goldman. 2006.
9. *Ethics of Spying: A Reader for the Intelligence Professional*, Volume 2, edited by Jan Goldman. 2010.
10. *A Woman's War: The Professional and Personal Journey of the Navy's First African American Female Intelligence Officer*, by Gail Harris. 2010.
11. *Handbook of Scientific Methods of Inquiry for Intelligence Analysis*, by Hank Prunckun. 2010.
12. *Handbook of Warning Intelligence: Assessing the Threat to National Security*, by Cynthia Grabo. 2010.

HANDBOOK OF WARNING INTELLIGENCE

Assessing the Threat to National Security

Cynthia Grabo

Jan Goldman
Series Editor

Scarecrow Professional Intelligence Education Series, No. 12

THE SCARECROW PRESS, INC.
Lanham • Toronto • Plymouth, UK
2010

Published by Scarecrow Press, Inc.
A wholly owned subsidiary of The Rowman & Littlefield Publishing Group, Inc.
4501 Forbes Boulevard, Suite 200, Lanham, Maryland 20706
http://www.scarecrowpress.com

Estover Road, Plymouth PL6 7PY, United Kingdom

British Library Cataloguing in Publication Information Available

Library of Congress Cataloging-in-Publication Data

Grabo, Cynthia M.
Handbook of warning intelligence : assessing the threat to national security / Cynthia Grabo, Jan Goldman.
p. cm. — (Scarecrow professional intelligence education series ; no. 12)
Includes bibliographical references.
ISBN 978-0-8108-7190-8 (cloth : alk. paper) — ISBN 978-0-8108-7166-3 (pbk. : alk. paper) — ISBN 978-0-8108-7095-6 (ebook)
1. Intelligence service—Methodology—Handbooks, manuals, etc. 2. Intelligence service—United States—Methodology—Handbooks, manuals, etc. 3. Intelligence service—United States—History—20th century. 4. Cold war. I. Goldman, Jan. II. Title.
JF1525.I6G73 2010
327.1273—dc22 2009043463

™ The paper used in this publication meets the minimum requirements of American National Standard for Information Sciences—Permanence of Paper for Printed Library Materials, ANSI/NISO Z39.48-1992.

Printed in the United States of America

Contents

Foreword

THE STORY BEHIND THIS BOOK began 40 years ago, and it is still incomplete. This is an unbelievable chronicle of perseverance and patience, which is both amazing and true.

The last time the author had seen her manuscript was in 1972. At the time, she was working as an intelligence analyst at the Defense Intelligence Agency in Washington, DC. As required by her agency, and all U.S. government employees in the intelligence community, she submitted the manuscript for approval before it could be published. Typically, government officials review manuscripts to make sure any information released cannot harm United States national security. Sometimes it is a matter of deleting a word or sentence throughout a manuscript before it can be publicly released. But not on this day. Cynthia Grabo's entire manuscript was stamped SECRET and confiscated, and just like that it became classified and the property of the U.S. government. Then, after 27 years, Cynthia decided it was time to retrieve her manuscript. She made numerous phone calls in an attempt to find her manuscript, but it was proving to be an almost impossible task. It was toward the end of her list on that April day in 1999 that she called the National Defense Intelligence College. The school receptionist transferred her call to me—because I happened to be the only faculty member still in his office eating lunch.

Although, we had never met, I admired Cynthia Grabo. I considered her one of the first intelligence analysts to understand the art and science of strategic warning, and I used many of her classified reports in my class. These reports taught students how to perform the methodology known as "indications and warning" intelligence. Her ability to write clear and concise intelligence reports was timeless. Nevertheless, I was shocked to get that phone call.

Working at Home

In January 1942, Cynthia was called into the dean's office at the University of Chicago. She was a student in the graduate program seeking a degree in international relations. A month earlier, the Japanese had bombed the U.S. Naval Base in Pearl Harbor, Hawaii. The United States was entering World War II, and the country needed everyone's support for the war effort. Cynthia was willing to help and the dean had recommended her for a position with the federal government. With assurance that she could still complete her degree requirements, she left school that month to begin work.

Cynthia arrived in Washington, DC, and was assigned to be a U.S. Army intelligence analyst. Her desk was in one of the several tents and buildings that sprang up around the city. A year later, Cynthia would soon become one of the first employees to work at the newly constructed War Department's building, known as the Pentagon.[1]

Cynthia was responsible for reading and writing classified intelligence reports. As an intelligence analyst, over the years she read and wrote intelligence reports during World War II, the Korean War, and the conflict in Vietnam. It was soon after the Soviet Union invaded Afghanistan that Cynthia decided it was time to retire. Overall, she had worked as an U.S. government intelligence analyst from January 1942 to January 1980. However, the Soviet invasion of Czechoslovakia in 1968 seemed to have the biggest impact on her and was the impulse behind her writing this book.

Cynthia was concerned with how the intelligence community analyzed and produced intelligence warning reports. She read the "raw intelligence"[2] depicting Soviet tanks as "likely" to enter and put down any form of democracy that the government of Czechoslovakia was willing to initiate and experiment with. However, not everyone saw this as a likely event and were completely surprised when the Soviet tanks crossed the borders for its attack. According to the U.S. State Department,

> The Warsaw Pact invasion of August 20–21 caught Czechoslovakia and much of the Western world by surprise. In anticipation of the invasion, the Soviet Union had moved troops from the Soviet Union, along with limited numbers of troops from Hungary, Poland, East Germany and Bulgaria into place by announcing Warsaw Pact military exercises. When these forces did invade, they swiftly took control of Prague, other major cities, and communication and transportation links.[3]

However, according to Cynthia,

> From the standpoint of history, it was a minor threat and for policy makers it was a minor blip on the screen, an enormously significant example of the Intelligence Communities (IC) "warning problem."[4] Most of the IC did not believe that the Soviets would do such a thing. This belief led to the rejection of the evidence. We had 6 to 7 analysts at the National Indications Center. I and one other member were the only two who believe the Soviets would invade. Therefore, the evidence to support the likelihood of an invasion was never considered. Signs pointing to an invasion included a letter from the Communist Party of the Soviet Union (CPSU) indicating that the CPSU would use 'all means necessary' to retain Soviet control over Czechoslovakia. French and Italian communist party leaders traveled to Moscow to plead with the Soviets not to invade. But the analysts didn't consider this evidence important . . . the analysts were the only ones. I think President Dubcek of Czechoslovakia himself was surprised. He didn't think the Soviets would really go through with it.[5]

Cynthia decided to write a book for newly hired intelligence analysts on how to read, write and analyze what she called "warning intelligence." It was one of the first educational books in professionalizing intelligence analysts, and this manuscript was both a textbook and a memoir. As a textbook, it taught analysts how to produce accurate reports and briefings, and as a memoir it highlighted the problems an analyst faced in getting "bad news" to the decision maker. At the time, Cynthia already had 25 years of experience in the intelligence community. From 1968 to 1971, she would return home from the Pentagon, and after dinner would sit down behind her typewriter to produce her book. When she completed the almost 700-page manuscript, one more "detail" needed to be completed. She would need to submit it through the Department of Defense security office. As a U.S. government employee in the intelligence community, then as now, Cynthia was obligated to submit all her writings for clearance and approval before they could be released for publication. Cynthia was aware of this requirement, and because there was nothing that could be considered classified, she was confident it would be approved for publication. As Cynthia points out in the "Author's Note,"

> This work is not an official project. It has not been written under the auspices of any agency, nor does it necessarily represent the views of any agency. It is being done entirely on my own time in addition to other duties, and all opinions expressed are my own.[6]

Unfortunately, this is where the process began to go awry. Government officials did not agree with Cynthia that this manuscript would not have a detrimental effect on U.S. national security. According to government officials, this publication had to be withheld from public dissemination. The book was classified SECRET and it became the property of the government.[7]

A Classified Government Publication

The manuscript was too important to sit on the shelf without being read; consequently, the government published 417 copies[8] of this recently "classified document" that Cynthia had written at home. For reasons that are unexplained, the government printing office published the manuscript as a trilogy. The three-volume textbook was titled *A Handbook of Warning Intelligence.* The first classified volume (chapters 1 to 18) was released in July 1972, the second volume (chapters 19 to 29) was released in November 1972, and the final volume (chapters 30 to 40) was released in June 1974. The books were distributed throughout the U.S. intelligence community. They became extremely popular, resulting in a second printing of the trilogy. The table of contents listed in this book is the actual listing of all the chapters (including the last 10 chapters, which remain classified and are not in this publication).

In 1980, after 38 years of working in the basement of the Pentagon, Cynthia completed her career in government service. As she prepared to retire, she wanted her book declassified. She never lost interest in having a commercial publisher distribute the book to a wider audience than the intelligence community. She wanted policy makers and all consumers of intelligence (including the public) to be aware of the arduous task of analyzing and producing intelligence reports. At her retirement ceremony, Cynthia received the prestigious National Intelligence Medal of Achievement; she did not, however, receive permission to have her manuscript declassified.

Thirty years after Cynthia wrote the book, and almost 20 years after she had retired and sought to get it declassified, she still wanted to get ownership of the manuscript. Thus in 1999 Cynthia decided to try again. She had a good reason to think the time was right to get her book returned. No longer could government officials claim the book needed to be kept classified because of the Communist threat. Most of the book deals with these countries and the threat they posed to United States foreign policy, and the Soviet Union no longer existed, along with the East European countries that were once under its domination in the Warsaw Pact. However, Cynthia ran into a new problem that went beyond fighting an enemy; it was finding a copy of her book. Her name recognition and copies of the book appeared to have evaporated along with the Berlin Wall.

Cynthia called all agencies in the intelligence community and she was running out of options and people. Finally, and probably out of final desperation, she called the Joint Military Intelligence College.[9] The secretary transferred her phone call to me. My office sits in a vault, officially known as a Sensitive Compartmentalized Intelligence Facility. I am surrounded with classified material. One of the unique features of this institution, as the only school accredited in the U.S. government with a degree in intelligence, is that you need to have a top-secret clearance to attend, along with the usual requirements of every other school.

Without any formality in her voice, Cynthia went directly to a short prepared introduction: "My name is Cynthia Grabo. I'm looking for a book and I was wondering if you could help me." Cynthia went on to describe the book on warning intelligence and that it had been almost 20 years since she last seen a copy of this government publication. According to her, it was time to get it declassified and published; but first she would have to find it. I leaned back in my chair to glance at my bookshelf. The metal structure contained dozens of books with the markings "Secret" or "Top Secret" on them. "Ms. Grabo, I think I have that book," I said.

The three-volume series sat between stacks of papers, reports and books. On the front of the publication of the first volume was stamped "copy 173 of 417 copies" and the word SECRET appears at the top and bottom of the thin orange cardboard cover.

In 1994, when I arrived at the college after an assignment at the Pentagon, several boxes sat in the corner of my new office. My first responsibility was to go through a dozen huge boxes filled with papers, books and journals. They were passed along to me from a departing faculty member and if I wanted the new office, I would need to go through the boxes. (Of course, anything I did not want to keep I would be responsible for destroying.) Most of the items were clearly outdated and rarely used, which was evident by the layer of dust on them. By the end of the day, many of the items were sent to the shredder to be destroyed. Cynthia's book, however, was invaluable for faculty members teaching intelligence analysis. I used parts of the book in my course.

Getting It Partially Declassified

Cynthia and I met the next day. Over lunch, it was apparent that she was both relieved and thankful that someone had a copy of her book, and I could not believe this was the woman whose intelligence reports on Korea and the Cuban Missile Crisis I used in my class. However, because Cynthia had retired from the federal government in 1980 and no longer had an active security clearance, we could not discuss the book. In a situation that was what can only be described as "Kafkaesque," she was no longer eligible to read the book that she had written.[10]

Soon thereafter, I requested and received a special clearance for Cynthia to come to the Defense Intelligence Agency to read her book. After reading it, she was completely convinced the book needed to be declassified and published. Over the next three years, I assisted with the partial success in getting the books declassified. Without reason, the government declassified only the first two volumes of the trilogy. The final volume, which appears in the table of contents of this publication, remains classified.

On weekends, Cynthia and I would extract the most important and relevant information into a readable and short book on how to perform warning intelligence. I knew that there were many books *about* intelligence, but few books on *how to do* intelligence. With a tone that is both instructional and encouraging, Cynthia's book had too much information for future generations of intelligence analysts. This became more apparent as the intelligence community expanded in a hiring binge after the attacks of September 11, 2001. In the end, 169 pages were culled from almost 450 pages of the original publication. The abridged version of the original manual represented the most essential information on the methodology of doing warning intelligence analysis. What we did not include in this abridged manuscript were many of the observations and insight from her years of experience.

The book, *Anticipating Surprise: Analysis for Strategic Warning*, was published by the Joint Military Intelligence College at the end of 2002. A thousand copies were published and within a few months the inventory was gone. Schools inside and outside of the government were using the book to teach intelligence analysis, while historians were learning more about how intelligence was produced and used during the cold war. Four years later, the University Press of America published the book with the same title.[11] Today, both books continue to be extremely popular, the book is cited in numerous articles and books, and it is considered essential reading on most intelligence and national security–related lists. Intelligence programs newly developed after 9/11 are using this book at the undergraduate and graduate levels of higher education. The book was nominated in 2005 for an award by the American Historical Association, and in 2008 its was translated into and published in Chinese.

The publication you are about to read is the unclassified and unabridged edition of Cynthia's original manuscript . . . up to a point. Missing is the third and final volume of the trilogy. That volume is still classified and under review for publication by the government. Nevertheless, the table of contents from that volume is being used because it appears in the other volumes. Since this is the full-length version, the decision was made to publish it under its original title, *Handbook of Warning Intelligence.* If and when the third volume is released by the government, it will be published.

Cynthia's objective when she began this project in 1968 was to publish the handbook exactly as she had written it. Although the final volume remains classified, it does not diminish the essential knowledge evident in this publication. There is a lot to learn from this book, and there is a lot of new intelligence personnel that needs the wisdom of "those who have been there and done that." Over the years, Cynthia has become a mentor to me and—more importantly—a friend. In late 2008, as she and I had lunch in a small café, the television on the wall showed another roadside attack on U.S. forces in Afghanistan. Without a doubt the world and the threat has changed since the cold war, but, understanding how to do good intelligence analysis has not changed . . . and, in Cynthia's book, you get to hear some pretty good stories.

Note: The views and opinions of the author do not reflect those of the U.S. government, Department of Defense, or any agency in the Intelligence Community.

Notes

1. The Pentagon, the largest office building in the country at that time, began construction on September 11, 1941, and was completed by January 15, 1943. The Pentagon covers 29 acres and contains 17.5 miles of corridors.
2. Information that has been collected but that has not been processed for validity.
3. U.S. State Department, "Soviet Invasion of Czechoslovakia, 1968." http://www.state.gov/r/pa/ho/time/ea/107190.htm (accessed August 1, 2009).
4. *Warning problem* is identified as a potential threat that when translated into threat scenario(s) postulate a sequence of events that, when this process is completed, represents an unambiguous threat.
5. Eggerz, Solveig. "Former DIA Analyst Writes Popular I&W Book," *Communiqué: A Defense Intelligence Agency Publication*, October 2003, vol. 15, no 6, 16–17.
6. Cynthia Grabo, *A Handbook of Warning Intelligence*, p. v.
7. A classification level is assigned to information owned by, produced by or for, or controlled by the United States government. Also, a designation assigned to specific elements of information based on the potential damage to national security if disclosed to unauthorized persons. The three levels in descending order of potential damage are Top Secret (unauthorized disclosure of which reasonably could be expected to cause *exceptionally grave damage* to the national security), Secret (information, the unauthorized disclosure of which reasonably could be expected to cause *serious damage* to the national security), and Confidential (information, the unauthorized disclosure of which reasonably could be expected to cause *damage* to the national security).
8. We know that only 417 books were published, because each copy was numbered and that number appeared on the cover.

9. The Joint Military Intelligence College is located in Washington, DC, and is part of the Defense Intelligence Agency. In 2007, the school changed its name to the National Defense Intelligence College.

10. Typically, when a person retires from the intelligence community, he or she will usually maintain access to their clearances for up to 5 years; unfortunately, it was 19 years since Cynthia had retired.

11. Cynthia Grabo, *Anticipating Surprise: Analysis for Strategic Warning*, Lanham, MD: University Press of America, 2006.

Author's Note to Original Edition

SOME YEARS AGO, a group from several intelligence agencies was discussing the question of indications analysis and strategic warning. Reminded by one individual present that analysts who used indications methodology had correctly forecast both the North Korean attack on South Korea and the Chinese intervention in Korea, a relative newcomer to the business said, "Yes, but you can't have done a very good job, because no one believed you." This bit of unintentional humor aptly describes much of the problem of warning intelligence. Why is it that "no one"—a slight but not great exaggeration—believes in the indications method, despite its demonstrably good record in these and other crises which have threatened our security interests? Can the reluctance to believe be in part for lack of understanding of the nature of indications analysis or lack of experience with live warning problems?

Within recent years, the effort to improve our so-called warning capabilities has concentrated heavily on problems of collection, crisis management, and automatic data processing. Relatively little attention has been paid to warning as an analytical problem, despite the fact that experience shows that some types of (if not identical) analytical problems do recur in crisis situations, and that there are lessons to be learned from a study of the past.

This handbook is the product of more than twenty years practical experience with indications and warning intelligence from the analytical standpoint. So far as I know, it is the first attempt to bring together a body of experience on the warning problem and to set forth some guidelines which may assist analysts, and others involved in the warning process, in the future. Although this work is intended primarily for desk analysts and their immediate supervisors, and for use in intelligence training courses, it is hoped that it will be of benefit also to higher level intelligence personnel, and to those at the policy level who are dependent on the intelligence process for strategic warning. For it is as essential, perhaps even more essential, that they understand the nature of warning intelligence as it is for working level personnel.

This work is not an official project. It has not been written under the auspices of any agency / nor does it necessarily represent the views of any agency. It is being done entirely on my own time in addition to other duties, and all opinions expressed are my own.

This volume is the second printing of the first four parts (Chapters 1 to 18). Both printings have been done by the Defense Intelligence Agency primarily for use within the military establishment at both intelligence and operational levels. Some distribution is being made to other intelligence agencies and government personnel who have expressed an interest.

There have been no substantive changes in the text for this second printing, only a few minor editorial changes. The pages of each chapter are numbered separately, but each page now also shows the chapter number (e.g., 5-9 signifies Chapter 5, page 9). The dates on the individual chapters are the dates when they were completed.

The remainder of this work (Parts V–IX, Chapters 19–40) is expected to be completed in 1973. It will probably be issued initially in two more volumes, the first in the fall of 1972 and the remainder the following year. Consideration is also being given to the preparation of an unclassified version, provided there is sufficient interest in it. Although the general principles and techniques of indications analysis and the table of contents are unclassified, a number of the examples used in this work—which give life and validity to the principles—must remain classified for an undetermined period.

I wish to express my thanks to several people for encouraging me to continue with this work, and for their comments. They include: Dr. Sherman Kent, former Director of the Office of National Estimates; Dr. Thomas G. Belden, Chief Historian, U.S. Air Force; the past and present Directors of the National Indications Center; the staff of the Defense Intelligence School; and other individuals in the Defense Intelligence Agency, particularly those in the Directorate for Intelligence who have arranged for the printing and distribution of this work. Above all, my thanks to two wonderfully efficient and cooperative secretaries, Misses Gertrude Trudeau and Kathleen Smith, who have typed this manuscript and thus made its publication possible.

Cynthia M. Grabo
National Indications Center
Pentagon
June 1972

1

General Nature of the Problem

The Function of Warning Intelligence

WARNING INTELLIGENCE, or as it is often called indications intelligence, is largely a post-World War II development. More specifically, it was a product of the early days of the cold war—when we began to perceive that the Soviet Union and other Communist nations were embarked on courses inimical to the interests and security of the free world and which could lead to surprise actions or open aggression by Communist states. Enemy actions in World War II—above all the Japanese attack on Pearl Harbor—also had dispelled many of the conventional or historical concepts of how wars begin. The fear that our enemies once again might undertake devastating, surprise military action—without prior declaration of war or other conventional warning—became very real. The advent of modern weapons and long-range delivery systems further increased the dangers and the consequent need for warning to avoid surprise attack.

Surprise military actions and even undeclared initiation of wars are not, of course, exclusively a modern phenomenon. History has recorded many such actions, going back at least to the introduction of the wooden horse into the ancient city of Troy. The military intelligence services of all major powers during World War II devoted much of their time to collection and analysis of information concerning the military plans and intentions of their enemies, as well as their capabilities, in an effort to anticipate future enemy actions. Many of the problems and techniques which we today associate with indications or warning analysis were recognized and practiced during World War II. The history of the war records some brilliant intelligence successes in anticipating enemy actions as well as such conspicuous failures as Pearl Harbor, the German attack on the Soviet Union and the Battle of the Bulge.

Thus, warning intelligence as we know it today really is not new. It is only that it has now become recognized as a distinct function of intelligence, as something which the intelligence system should be examining full time, in war and in peace, so long as there are admittedly hostile countries with powerful military forces which could be employed against us or our allies. Warning intelligence today—although closely allied with current intelligence, estimates and basic intelligence—is nonetheless somewhat separate. Its function is to anticipate, insofar as collection and analysis will permit, what the enemy is likely to do, and particularly whether hostile nations are preparing to initiate military action in the foreseeable future. Its role is to provide the policy maker and the military commander with the best possible—and earliest possible—judgment that some action inimical to the security of the nation may be impending.

Now there are many things that can occur in the world which could be inimical to U.S. interests and which might entail some U.S. action, either political or military. A coup in Bolivia, the outbreak of civil war in the Congo, the kidnapping of a diplomat in Guatemala, conflict between Pakistan and India, new Soviet military aid shipments to Algeria, the assassination of a pro-Western chief of state anywhere—all these and a host of other military and political developments are of concern to the U.S., and it is a function of intelligence, insofar as possible, to anticipate such developments and to alert the policy maker to them. In this sense, "warning" could be said to be an almost unlimited responsibility of the intelligence system and to involve potentially almost any development anywhere in the world. The current intelligence process is involved daily with this kind of problem. Throughout the system, at both the national level and overseas, operational centers, alert centers and watch offices are concerned day and night with keeping up with problems and potential crises of this type.

For warning intelligence also to attempt to deal with an unlimited number of problems and crises of this nature would obviously result in an unnecessary and expensive duplication of effort. The warning effort would become largely another current intelligence system if it attempted to cope with every or even most potential crises or unexpected developments which might be of concern to the policy maker or commander. Still worse, the warning effort could become so diffused and distracted with secondary problems that its primary function could be impaired.

The term "warning intelligence"—as it has usually been used since World War II and as it will be used throughout this book—is generally restricted to:

(a) direct action by hostile Communist states against the U.S. or its allies, involving the commitment of their regular or irregular armed forces; (b) other major developments affecting U.S. security interests in which such hostile Communist states are or might become directly involved; and (c) military action by Communist states against other nations (not allied with the U.S.), including action against other Communist states.

Obviously, no absolute guidelines or directives can be laid down in advance as to when, or under what circumstances, any particular situation or area properly becomes a subject for analysis or judgments by warning intelligence. Some situations are demonstrably warning problems most gravely threatening U.S. forces or security interests, such as the Chinese intervention in Korea or the Cuban missile crisis. Other situations, although not involving such direct threats of confrontation for the U.S., nonetheless pose grave risks of escalation or involvement of other powers; a series of Middle East conflicts and crises have been in this category. Some areas—for many years Berlin, and almost continually Southeast Asia—have been long-term, almost chronic, warning problems; while others, such as the Taiwan Strait, may occasionally become critical subjects for warning intelligence. Conflicts or potential conflicts between Communist states—the Soviet suppression of the Hungarian revolt, the invasion of Czechoslovakia, and the Sino-Soviet border—also have been subjects for warning analysis and judgments. And a number of lesser crises, or threats of possible crises or actions, have been examined over a period of years, in order to provide judgments for higher officials.

Whether or not an immediate crisis or threat exists, however, the function of warning intelligence also is to examine continually—and to report periodically, or daily if necessary—any developments which could indicate that a hostile nation is preparing, or could be preparing, some new action which might be inimical to U.S. security interests. It examines developments, actions, or reports of possible actions or plans—military, political and economic—throughout the Communist world, and in other nations when pertinent, which might provide any clue to possible changes in policy or preparations for some future hostile action. It renders a judgment—positive, negative, or qualified—that there is or is not a threat of new military action or an impending change in the nature of ongoing military actions of which the policy maker should be warned. And it usually includes as well some analytical discussion of the evidence at hand to support the conclusions.

Thus warning intelligence serves both a continuing "routine" function and an exceptional crisis function. Routinely, from day to day or week to week, its report may say little and its final judgment may be negative. Yet it serves as a kind of insurance that all indications or possible indications are being examined, discussed and evaluated, and hopefully that nothing of potential indications significance has been overlooked. In time of crisis or potential crisis, it serves (or should serve) the function for which it really exists—to provide warning, as clear and unequivocally as possible, of what the enemy is probably preparing to do.

U.S. Warning Intelligence: A Synopsis

A full account of the history of U.S. warning intelligence and how it evolved would consume many pages, and would be superfluous for the purposes of this study. A brief discussion of how the system has developed, however, may be useful for the student of indications methodology.

There is a story that President Truman—an avid reader of intelligence reports—was concerned with the quantities of conflicting reports and rumors concerning possible Soviet plans and intentions at the start of the Berlin blockade in 1948. "Who," he is said to have asked, "is keeping track of all these indications?" Whether this is true or not, and if so whether his remark contributed directly to the U.S. indications effort, the genesis of that effort can be traced to about this time. The Berlin blockade, together with other signs that the USSR and the Chinese Communists were on an expansionist course, inspired an effort in U.S. intelligence agencies, as well as by British intelligence, to bring together the various fragments of information or indications of possible Soviet plans and intentions. By early 1949, at least two U.S. indications groups were at work extracting and compiling information relating to Communist military and political activities and plans both in Europe and Asia. Both groups—one of which met periodically on an inter-agency basis under CIA auspices, and the other in Army Intelligence—produced very informal periodic reports which received a limited distribution and were probably never seen by policy makers. This was most unfortunate, since both groups were gravely concerned with the growing number of indications by the spring of 1950 that the Communist Bloc was getting ready for aggressive action in one or more areas. Across the board, the number of indications was impressive and the outlook ominous.

The North Korean attack on South Korea on 25 June 1950 (which some analysts believe would have been only the first of several aggressive actions, had we not intervened) immediately prompted an expanded and more formal indications effort on an inter-agency basis. (It may be hard for those accustomed to intelligence as it operates today to believe it, but it is a fact that at that time there was virtually no interagency research collaboration and no formal coordinated reporting.) Within weeks, representatives from Air and

Navy were attending a weekly meeting under the Army's chairmanship, which became known as the Joint Intelligence Indications Committee (JIIC). Shortly later, representatives from the Department of State, CIA, the AEC and the FBI also joined the group. As of January 1951, it officially became the Watch Committee under the auspices of the Intelligence Advisory Committee (the predecessor of the United States Intelligence Board). The weekly reports of the U.S. Watch Committee have been numbered serially since the formation of the JIIC in August 1950. From 1950 until 1954, the research and secretarial backup for the committee was provided by the Army, with some assistance from Air and Navy.

In 1954, following extended study, the indications effort was reorganized and expanded to provide an inter-agency working staff for the Watch Committee, known as the National Indications Center (NIC), and a senior CIA official assumed the permanent chairmanship of the Watch Committee. In 1958, NSA formally became a member of the Watch Committee. In 1961, DIA replaced the service representatives on the Committee and the latter became observers. Otherwise, there has been little change in the composition or function of the Watch Committee since its establishment. The NIC remains the only full-time inter-agency staff devoted solely to intelligence analysis and reporting.

Probably more important than these formal changes has been the growing recognition of the importance of warning and indications intelligence throughout the intelligence system. Nearly all intelligence agencies today have some personnel or effort devoted exclusively or primarily to the indications effort. In addition, indications centers have been established in virtually all U.S. military commands under the auspices of DIA, and many of these centers have their own Watch meetings and disseminate weekly Watch Reports devoted to the particular problems of their area. Counterparts of the U.S. Watch Committee also exist in the British Commonwealth countries (although their functions are somewhat broader), and the weekly reports of these various committees are exchanged.

Further, intelligence analysts in all agencies have a part in the indications effort, either through the contribution of items, briefing of their Watch representatives, attendance at Watch meetings, or review of the draft reports. Indicator lists, once informally prepared, are now extensively staffed and coordinated with specialists in the various agencies (see Chapter 7, Indicator Lists). Tripartite indicator lists (mutually agreed by the U.S., UK and Canada) are also extensively staffed and periodically reviewed.

There is no question today of the status of warning intelligence and that the inter-agency indications effort, from its modest beginnings in 1948–49, is probably here to stay. From almost no coordination, we have come to almost total coordination. If there is any danger to the system today, it is not that indications intelligence is not recognized, but that excessive formality or emphasis on coordination and unanimity can lead to watered down judgments or suppression of minority views. These problems will be discussed in later chapters.

Major Warning Problems of Recent Years

Since 1946, we have been confronted repeatedly with some crisis or threatened crisis which either has or might have involved the use of force by a Communist power. In the interval between 1953 and 1965 when U.S. forces were not directly involved in a war in Asia, a series of threats in Berlin, the Middle East, the Taiwan Strait, Vietnam, Laos, and Cuba provided a continuing series of problems to keep the warning system on its toes. No indications analyst was bored.

To refresh the mind of the reader and since reference to these various crises will be made in ensuing pages of this work, it may be well to identify the major problems which have been of concern to warning analysts since 1950. We will consider them here not in chronological order, or even in order of their importance to U.S. policy officials or military leaders, but rather grouped roughly in terms of their significance as warning problems.

True, Major Warning Problems

In this category are the great classic warning problems to which the student of indications will wish to devote his major attention. It will be noted that only one of the four below involved a commitment of U.S. forces, although a second (Cuba) was possibly our greatest potential threat since World War II. All four of these examples were marked by massive military preparations, secret but also obvious deployments of forces, some attempts at deception, a wide variety of political indications, and (except in the case of Cuba where it was narrowly averted) by ultimate military conflict. They are:

- The North Korean attack on South Korea—June 1950—and the Chinese Communist intervention in Korea—October–November 1950
- The Hungarian revolt and the Suez crisis (simultaneous but separate crises)—October–November 1956
- The Cuban missile crisis—October 1962
- The invasion of Czechoslovakia—August 1968

Long-Term, Continuing Warning Problems

In this category are the chronic problems which have plagued the nation year after year. While they have for the most part lacked the drama of the major crises noted above and in one instance (Berlin) did not lead to conflict, they nonetheless hold many useful lessons for the warning analyst. All three have also been marked by periods of acute crisis for warning intelligence. They are:

- Berlin: a continuing problem from World War II, a critical problem in 1948–49 and from 1958–62, and a full-blown crisis in the summer of 1961
- Vietnam: a problem since 1950, with crises in 1954, 1965 and 1968; a persistent indications as well as military problem
- Laos: with Vietnam, a problem since 1950, with several critical periods such as 1960, 1962 and 1964

Lesser but Important Crises, or Potential Crises

In this category we have placed most of the remaining major problems which have confronted the indications effort in recent years. They have little in common, but all involved either the threat of force or actual employment of it. They are listed chronologically:

- The Turkish-Syrian crisis—summer 1957
- The Lebanese crisis—July 1958
- The Taiwan Strait crisis—August 1958
- The Sino-Indian border conflict—October 1962
- The Arab-Israeli six-day war—June 1967
- The Sino-Soviet border—a continuing problem but especially spring–summer 1969

The reader may have noted that, despite the number of crises listed above (and other lesser problems could be added), significant warning problems really have not been as frequent as some believe. While we have had continuing critical national problems, there are only a dozen or so in the past 20 years which merit study as indications problems, and fewer than that which are warning "classics" of enduring value for the analyst. Moreover, these crises have covered a variety of areas extending all around the periphery of the Communist world and to Cuba. They have involved North Korean, Chinese, Vietnamese, Laotian, Arab, Israeli, Soviet and Warsaw Pact forces. Few analysts, even those who have remained on one area for an extended period, have experienced as live problems more than two or three of these crises. Although involving widely separated areas, some of these problems have had considerable similarity as warning problems. For example, the Chinese intervention in Korea and the Soviet invasion of Czechoslovakia are remarkably alike as analytical problems, yet very few analysts were actively involved in both.

This relative infrequency of true warning problems—the lessons most valuable for the training of indications analysts—is one of the major difficulties for the warning system. Unlike numerous other types of intelligence analysis on which there is a continuing flow of live and pertinent material (order -o tile, weapons production, budgets, etc.), an analyst may experience a live warning problem only once or twice in his career. He may be trained on his first crisis and no real personal experience or background to bring to it.

This handbook has been written to help bring some of this experience to analysts who have never worked a true, live warning problem. While there is no adequate substitute for the live experience, it is hoped that the succeeding chapters will provide the analyst vicariously with some of the experience he might gain from the real thing.

2

Definitions of Terms and Their Usage

UNLIKE MANY SPECIALIZED TYPES of knowledge, warning intelligence has developed very few occult or technical terms. For the most part, it uses the same terminology as current or estimative intelligence, and is designed to be intelligible to "lay readers" not familiar with technical or scientific terms as well as to military experts.

There are a number of terms frequently used in warning intelligence, however, with which any writer or reader of the product will soon become familiar and which have acquired something of a specialized meaning. Some of these terms are in "accepted" usage, in that they appear in such volumes as the *JCS Dictionary of US Military Terms for Joint Usage* others have a generally well understood or accepted meaning. Some, on the other hand, have never been precisely defined and have been subject to somewhat varying interpretations. Thus, the following "definitions" cannot be described as graven in stone but are only the most usually accepted usage in the warning field.

Indications and Indications Intelligence

The term "indications intelligence" is considered synonymous with "warning intelligence" and is concerned with those pieces of information which relate to what the enemy is or possibly is preparing to do. The *JCS Dictionary* definition is: "Information in various degrees of evaluation, all of which bears on the intention of a potential enemy to adopt or reject a course of action." By common dictionary definition, the word "indicate" connotes something less than certainty; an "indication" is a sign, a symptom, a suggestion, a ground for inferring, a basis for believing, or the like. Thus, the choice of this term to denote the nature of warning intelligence was a realistic recognition that warning itself is likely to be less than certain and to be based on information which is incomplete, of uncertain evaluation or difficult to interpret.

An *indication* can be a development of almost any kind. Specifically, it may be a confirmed fact, a possible fact, an absence of something, a fragment of information, an observation, a photograph, a propaganda broadcast, a diplomatic note, a call up of reservists, a deployment of forces, a military alert, an agent report, or a hundred other things. The sole proviso is that it provide some insight, or seem to provide some insight, into the enemy's likely course of action. An indication can be positive, negative, or ambiguous (uncertain). For example, a continuing buildup of Soviet troops along the Chinese border is a positive indication of Soviet military preparations for possible conflict with China. A lack of evidence of concurrent mobilization or extraordinary requisitioning of transport is (probably) a negative indication that any action is imminent. A simultaneous offer to negotiate is ambiguous or uncertain as an indication—it may be a genuine effort to avert conflict but it could be deception or a device to gain time. Uncertainty concerning the meaning of a confirmed development is usually conveyed by such phrases as: it is a possible indication, it may indicate, it suggests, etc. Uncertainty as to the validity of the information itself may also be phrased this way, but more accurately this is conveyed by a phrase such as: if true, this indicates, etc.

Developments may even be "indications of indications" rather than positive signs that something abnormal has occurred. The issuance of new cards to reservists may indicate that they are to be called up; this may in turn indicate that a partial mobilization is planned, or it may indicate only that they will be called up for routine refresher training. Or the whole thing may indicate nothing except some procedural change in mobilization plans which required the issuance of new cards.

An *indicator* is a known or theoretical, step which the enemy should or may take in preparation for hostilities. It is something which we anticipate may occur, and which we therefore usually incorporate into a list of things to be watched for, known as "indicator lists" (see Chapter 7). (Information that any of these steps is actually being implemented constitutes an *indication*.) This distinction in terms—between expectation and actuality, or between theory and current developments—is a useful one, and those in the warning trade have tried to ensure that this distinction between indicators and indications is maintained. Many fail to make this careful distinction, however, and the word "indicator" very frequently is used to mean an actual development which has occurred, or an indication. No one, however, refers to "indicators centers" or "indicators intelligence"—an indication that the difference in meaning is actually generally understood.

Strategic versus Tactical Warning

The term *strategic warning* somewhat regrettably has no single, accepted definition in US intelligence and operational usage. To those in the field of warning intelligence, strategic warning is generally viewed as relatively long-term, or more or less synonymous with the "earliest possible warning" which the warning system is supposed to provide. Thus, in this meaning, strategic warning hopefully can be issued weeks (or even months) in advance, when a large-scale deployment of forces is under way, or the enemy has made known his political commitment to some course of action entailing the use of force. This judgment—of the probability of military action at some time in the future—is unrelated to the imminence of action. The judgment may be possible only when enemy action is in fact imminent, but it may also be possible long before that. In this sense—and it is the sense in which intelligence normally uses it—*strategic warning* is an assessment that the enemy has or probably has taken a decision to employ force. When he may implement that decision is another matter. The policy maker needs the warning as early as possible.

In operational military use, however, the term strategic warning has a considerably different meaning and is defined in the *JCS Dictionary* as: "A notification that enemy-initiated hostilities may be imminent. This notification may be received from minutes to hours, to days, or longer, prior to the initiation of hostilities." Thus, in this usage, the policy or operational level anticipates a definite "notification" of some type (either through intelligence or operational channels) that enemy attack probably is impending in the very near future. This strategic warning then entails decisions by the highest levels of government—the interval between the receipt of strategic warning and the outbreak of hostilities is divided into "strategic warning predecision time" and "strategic warning post-decision time."

As the reader proceeds with this book, he may come to perceive some of the unreality in these definitions. It is, for example, most unlikely that decisions would be deferred until enemy action appeared imminent, provided evidence of preparation for such action had previously been available. In practice, it is not the recognition of imminence of action but the recognition of probability of action that constitutes strategic warning. For the intelligence warning system, strategic warning is likely to remain a judgment of likelihood of action rather than of imminence of such action, regardless of how terms may be defined. It is important, however, that the intelligence system recognize that the term strategic warning also has a very specific operational usage.

Tactical warning is much more easily defined, although even here there is some shading in meaning. Strictly defined, tactical warning is not a function of intelligence (at least not at the national level) but is an operational problem. It is that warning that would be available to the commander on the front line or through the radar system or other sensors, which would indicate that the attacking forces were already in motion toward the target. The *JCS Dictionary* defines tactical warning as: "A notification that the enemy has initiated hostilities. Such warning may be received any time from the launching of the attack until it reaches its target." Under this definition, it is obviously too late for the intelligence system to come to any judgments to convey to the policy maker or commander; all evaluation as well as response can only be in the hands of the operational level. In practice, the line between strategic and tactical warning may not be so precise. In fact, many would consider that intelligence had issued *tactical warning* if it were to issue a judgment that an attack was imminent, particularly if this followed a series of earlier judgments that attack was likely. When would warning of the Tet Offensive of 1968 have become tactical? Would a warning issued in the early afternoon of 20 August 1968 of a Soviet invasion of Czechoslovakia have been strategic or tactical?

For the intelligence warning system, there is one simple rule: Its function is *strategic*, in the true sense of the term. If it cannot issue a warning until minutes before, whatever adjective is attached to that warning, it will have failed in its job. Whatever word is applied to it, warning must be given far enough ahead to permit the policy maker to make some considered decisions and for the military commander to take some essential preparedness measures.

Definitions of Timing

It is a symptom of the difficulty of ascertaining when, as well as whether, the enemy is preparing to attack or to take some other action that so many different phrases are in use to attempt to describe the when of it. Even negative judgments are plagued by this problem: shall we say that hostile action is not imminent, will not occur in the near future, foreseeable future or what? No clear-cut or accepted definitions of any such terms have ever been formally agreed on, and the following are offered only as general guidelines to the usage over the past twenty years or so at the national intelligence level:

Imminent: Used positively (attack is imminent), the term connotes an immediacy *and probability* of action that brook no delay—from right now to probably a few days at the most. Used negatively (attack is not imminent), a longer time frame is usually implied, perhaps several days to a week; we mean that we don't see this right now, but we would not be surprised to have it happen before long.

Immediate future: A frequently used term, usually in the negative (we do not expect Arab-Israeli hostilities in the immediate future), it has usually been interpreted to mean about a week, more or less. However, much depends on context or circumstances. If reached in a weekly report, it often is intended to cover the interval until the next report.

Near future: Somewhat longer than immediate future, but of very imprecise duration. Near future can convey anything from a week or a few weeks up to several months, or even can be used to mean foreseeable future. Used in the negative (we conclude that the USSR will not initiate hostilities in the near future), it has long been standard "warning conclusion" terminology.

Soon: Another indefinite term, usually about equivalent to "near future."

Foreseeable future: Also indefinite, but presumably in the near future, this term really is intended to convey a greater sense of certainty about things, often without regard to time. "We do not expect a change in Chinese hostility toward the US in the foreseeable future" conveys a sense that they are really pretty set in their ways and that we consider that a change is unlikely for some time. If we were less certain, we would select some other phrase conveying more doubt or a shorter time frame. In fact, many conclusions phrased in language of shorter time frame (such as near future) really mean as far ahead as we can see now. We just want to leave some doubts about how far ahead we can see or how certain we are that we know very much. A prudent precaution.

3

What Warning Is and Is Not

BEFORE WE ENTER INTO A DETAILED DISCUSSION of the warning process and the analytical techniques involved, it is necessary to dispel some widespread misconceptions about what warning is and is not. We shall look first at—

What It Is Not

a. Warning is not a commodity. It is not something which the analyst, the intelligence community, the policy maker or the nation has or does not have. This frequent misconception is expressed in casual questions such as, "Did we have warning of the Soviet invasion of Czechoslovakia?" Particularly when taken out of context, this misconception has also appeared in high-level official documents. Thus, one of several years ago said, inter alia, "It cannot be concluded that the US surely will, or surely will not, have strategic warning."

Warning is not a fact, a tangible substance, a certainty, or a provable hypothesis. It is not something which the finest collection system should be expected to produce full-blown or something which can be delivered to the policy maker with the statement, "Here it is. We have it now."

Warning is an intangible, an abstraction, a theory, a deduction, a perception, a belief. It is the product of reasoning or of logic, a hypothesis whose validity can be neither confirmed nor refuted until it is too late. Like other ideas, particularly new or complex ideas, it will be perceived with varying degrees of understanding or certitude by each individual, depending on a host of variable factors. Among them may be: his knowledge of the facts behind the hypothesis, his willingness to listen or to try to understand, his preconceptions of what is a likely course of action by the enemy, his cognizance of that nation's objectives and its military and political doctrine, his knowledge of history or precedent, his objectivity, his imagination, his willingness to take risks, other demands on his time, the attitude of his superior, his confidence in the man who briefs him, his dependence on others, his health, or even what he had for breakfast and whether he quarreled with his wife. These and other factors may influence his receptivity to the idea.

There are times, of course, when the volume of provable evidence, the sheer number of facts pointing to the likelihood of war is so overwhelming and so widely recognized that the conclusion is almost inescapable. This was the situation which prevailed before the outbreak of World War II and particularly in the week immediately preceding it, when the man on the street if he had read the newspapers probably was as qualified to judge that war was imminent as were the heads of state. In such a case, everybody, in the strategic if not the tactical sense, "has warning."

On the other hand, it is readily apparent that virtually no one—and particularly not the people who mattered—"had warnings" of the Japanese attack on Pearl Harbor, although the buildup of Japanese military power and Tokyo's aggressive designs in the Pacific were almost as well-recognized as were Hitler's designs in Europe.

No one can say what situation might prevail on the eve of World War III. We can say that in a series of lesser conflicts and crises since World War II warning rarely has been as evident as in September 1939. The odds therefore are that warning will remain an uncertainty in the years to come.

b. Warning is not current intelligence. This opinion will no doubt surprise a lot of people who have come to look on warning as a natural by-product or hand maiden of current analysis. Who is better qualified to detect and report indications of possible impending hostilities than the military and political analysts whose function is to keep on top of each new development and fast-breaking event? Is it not the latest information which we most need to know in order to "have warning"?

The answer to this is both yes and no. It is imperative that the analyst receive promptly and keep track of each new piece of information which may be an indication of hostilities. The analysts who are to produce the intelligence needed for warning cannot afford to fall behind with the incoming flood of paper lest they miss some critically important item. Both the collection and the processing of information must be as current as possible, for we know of instances in which the delay in receipt of information by even a few hours has contributed to serious failures to have drawn the right conclusions.

Nonetheless, the best warning analysis does not flow inevitably or even usually from the most methodical and diligent review of current information. The best warning analysis is the product of a detailed and continuing review in depth of all information going back for weeks and months which may be relevant to the current situation, and of a sound basic understanding of the potential enemy's objectives, doctrines, practices and organization.

The latest information, however necessary it may be to examine it, will often not be the most useful or pertinent to the warning assessment. Or, if it is, it may only be because it can be related to a piece of information received weeks before, or because it may serve to confirm a hitherto uncertain but vital fragment of intelligence which the analyst has been holding for months. Only in rare instances where events erupt very suddenly (e.g., the Hungarian revolt in 1956) can indications or warning analysis be considered more or less synonymous with current analysis. Most crises have roots going deep into the past, much farther than we usually realize until after they erupt. Preparation for war or possible war often can be traced back for months once it becomes clear that a real threat exists, and pieces of information which appeared questionable, unreliable or even ridiculous when received will suddenly have great relevance to the present situation, provided the analyst has saved them and can fit them into the current pattern. Further, information which is months old when it is received (and therefore scarcely rates as current intelligence) may be immensely valuable. An indication is not useless or invalid because it occurred months ago but you just found out about it today; it may help to demonstrate that the preparations for conflict have been far more extensive and significant than you had believed.

In normal times, the current analyst must cope with a large volume of paper. In times of crisis, he may be overwhelmed, not only with lots more paper but with greatly increased demands from his superiors for briefings, analyses, crash estimates and the like. It is no wonder in these circumstances that he can rarely focus his attention on the information which he received last month or find the time to reexamine a host of half-forgotten items which might be useful to his current assessment. The night duty officer, who may be the most "current" of all, will likely as not never even have seen many of the items from the preceding months which might be important now.

In addition, it may be noted that the weeks or days immediately preceding the deliberate or "surprise" initiation of hostilities may be marked by fewer indications of such action than was the earlier period. Or, as it is sometimes expressed, "warning ran out for us" ten days or two weeks before the attack occurred. (The reasons for this seeming lull before the storm, and how to cope with it, are discussed in Chapter 28.) Given this circumstance, the strictly current intelligence approach to the problem can be misleading or even dangerous. Since there are not many new indications to report, the current intelligence publication may convey the impression (and the analyst may actually come to believe) that the threat is somehow lessened, when in fact it is being maintained and may be increasing.

It is important to recognize that, in time of approaching crisis when many abnormal developments are likely to be occurring, the current intelligence analysts more than ever will need the assistance of those with detailed expertise in basic military subjects, such as mobilization, unit histories, logistics, doctrine, a variety of technical subjects and other topics. Even such things as the understanding of terminology rarely if ever noted before may be of vital importance. It is a time when current and basic intelligence must be closely integrated lest some significant information be overlooked or incorrectly evaluated.

There is still another difference to be noted between the current intelligence and warning processes, and that is the nature and content of the reporting. Since the primary function of the current analyst is not to write warning intelligence but to produce good current items, he will necessarily have to omit from his daily reporting a large number of indications or potential indications. There may be a variety of reasons for this, such as: the indications are individually too tenuous or contradictory, some of them are not current, there are just too many to report them all, they don't make a good current story, and a number of them (sometimes the most important) are too classified for the usual current intelligence publication or the analyst is otherwise restricted from using them. Some persons looking into the warning process or attempting to reconstruct what happened in a crisis have initially believed that the best place to look is in current intelligence publications. While there is nothing wrong with this approach to a point, and providing the situation which prevailed at the time is understood, no one should expect to find the whole story or the real story of what was happening in these publications. To pick an extreme example, in the week between the discovery of Soviet strategic missiles in Cuba on 15 October 1962 and President Kennedy's speech announcing their discovery on 22 October, the intelligence community was precluded in all its usual publications from alluding to the discovery at all, and for obvious reasons.

The foregoing are some of the ways in which warning and current analysis and reporting are distinguished from one another—or should be. It is in part for these reasons that it has been deemed prudent and desirable to have indications or warning specialists who, hopefully, will not be burdened or distracted by the competing demands placed on current analysts and will be able to focus their attention solely on potential indications and their analysis in depth. To accomplish this, these analysts must also recognize the value of basic intelligence and the importance of understanding how the enemy goes about preparing for war. (This aspect of the problem is discussed at length in Part IV.)

c. Warning is not a compilation of "facts." Nor is it a compilation of possible or potential facts or indications, however useful these may be. It is certainly not the intention of this study to downplay in any way the importance of a diligent and imaginative pursuit of indications. They are the foundation on which warning is built. The more hard evidence which is available, the more indications of high validity and significance, the more indisputable facts which can be assembled, the more likely it is—at least in theory—that there will be warning or that we shall come to the right conclusions. It all sounds so simple, until we are confronted with a condition and not a theory. In retrospect, it all looks so clear.

In actuality, the compilation and presentation of facts and indications is only one step in the warning process, occupying somewhat the same place that the presentation of testimony in the courtroom does to the decision of the jury and ruling of the judge. Just as the seemingly most solid cases backed by the most evidence do not necessarily lead to convictions, neither do the most voluminous and best documented lists of indications necessarily lead to warning. On the other hand, a very few facts or indications in some cases have been sufficient to convince enough people, or more importantly the right people, of an impending grave danger of hostilities or other critical actions. Some facts obviously will carry a lot more weight than other facts. The total number of different facts, or reported or possible facts, will not be as important as the interpretations attached to them. Too many facts or indications may even be suspect—why should it be so obvious? There must be something more (or less) than meets the eye.

A major portion of this treatise is devoted to a discussion of the relationship of facts and indications to warning, so these problems will not be elaborated here. Using the analogy with the courtroom case, however, we may note a number of possible reasons why the mere presentation of facts or evidence or statements of eyewitnesses may not produce a convincing case for the jury or a conviction:

- The statements of the prosecution's key witness are disputed by other witnesses.
- An eyewitness is demonstrated to be of uncertain reliability or an otherwise questionable reporter of the "facts."
- Several other important witnesses have not been located and therefore cannot be interrogated.
- The evidence, although considerable, is largely circumstantial.

- The defendant has a pretty good reputation, particularly in recent years, and has not done such a thing before.
- There appears to be no clear motive for the crime.
- The jury, despite careful screening, has been influenced by prior coverage of the case in the newspapers.
- The judge, a strict constructionist, rules that certain important material is inadmissible as evidence.

d. Warning is not a majority consensus. Both because of the importance of warning and because the process incidentally will usually involve analysts (and their supervisors) in a variety of fields, a large number of people are likely to participate in some aspect of the production of warning intelligence. Insofar as various specialists are called upon to present their expert opinions or knowledge on various problems of fact and interpretation (What is the maximum capacity of this railroad? What is the estimated TO&E of the Soviet tank division?), this is obviously all to the good. The intelligence community and the policy maker need every bit of expert help they can get on such questions. No one would presume to be able to get along without assistance of this type.

It does not necessarily follow, however, that the more people introduced into the warning process the better the *judgment* is going to be. Experience has shown that a consensus of all the individuals who have contributed something to the analysis of the problem, together with their supervisors, those responsible for making estimates, and others who may have an interest is not more likely to be correct than the judgments of analysts who have had experience with other warning problems and who are on top of all the information which is available in the current situation. Quite often the effect of bringing more people into the judgment process is to dilute the judgment in the interests of compromise and unanimity.

Lamentable as it may be, the fact is that the most nearly correct judgments in crisis situations over a period of years often have been reached by a minority of individuals. This does not usually mean that one agency is right and others are wrong—it is not that political analysts will come to right conclusions and military analysts to wrong conclusions, or vice versa. What usually happens is that a majority in all agencies is wrong—or at least not right. Thus the situation is not taken care of by the usual device of a dissenting agency footnote, since it will be a minority in each agency (not a majority in one) which will be in dissent.

Obviously, no one should say that because this situation has prevailed in so many cases from Pearl Harbor to the invasion of Czechoslovakia that it will always be so. (It is the hope that this treatise in some small measure will help to remedy this and bring more persons in the intelligence community to an understanding of the nature of the warning process and the problems of analysis and interpretation which must be faced.) It is important, however, that those involved (and particularly policy makers) understand that this has been the case in the past and that it is enormously important to the warning process that the views of the qualified and experienced minority be given an adequate hearing.

What Warning Is

a. Warning is an exhaustive research effort. Although, as we have noted, warning will not flow simply or inevitably from a compilation of facts and indications, it is also imperative to the process that the facts, potential or possible facts, and other indications be most diligently and meticulously compiled and analyzed. *It is impossible to overemphasize the importance of exhaustive research for warning.* It is the history of every great warning crisis that the post-mortems have turned up numerous, sometimes dozens, of relevant facts or pieces of information which were available but which, for one reason or another, were not considered in making assessments at the time. This is apart from the information in nearly every crisis which arrives just a little too late and which in some cases could have been available in time provided the substantive analysts, their supervisors, or their counterparts in the collection system had recognized a little earlier the importance of following up certain facts or leads. (See Chapter 4, Warning and Collection.)

All those associated with the warning business in any way—from the lowest ranking analyst to the policy maker—should beware first and foremost of the individual who is reaching his judgments on a likely course of enemy action based on something less than the most detailed review of the available evidence which may be directly or indirectly related. While this advice may seem so elementary that it may be taken for granted, the fact is that it cannot be taken for granted in crisis situations. In actuality, experience has shown that a large number of individuals—and often including those whose judgments or statements will carry the most weight—are rendering opinions in critical situations either in ignorance of important facts or without recognizing the relevance or significance of certain information which they may know. Indications and current analysts have been simply appalled, usually after a crisis has broken, to discover how many persons only slightly higher in the intelligence hierarchy were totally unaware of some of the most critical information from a warning standpoint. And warning analysts have been almost equally chagrined and remorseful to look back at the information which they themselves had available or could have obtained but which they just did not make use of or fit into the picture correctly.

How does this happen? How can it be that the great machinery of US intelligence, which is capable of spectacular collection and comprehensive analysis on many subjects, can fail to deal with all the facts or to carry out the necessary research in a warning situation? The answers to this are complex and some of the factors which contribute to this problem are dealt with at length in succeeding chapters. We will therefore note here only two of the more obvious difficulties which arise and which may impede the research effort and the surfacing of all the relevant facts:

The intelligence research system is set up primarily to analyze certain types of information which have recognized status as intelligence "disciplines" and on which there is a more or less continuing flow of material—order of battle, economic production, weapons developments, foreign policy to name a few. In crisis situations, great volumes of new material, some of it of a nature not usually received, may suddenly be poured into the system. In order to cope with this, agencies often set up special task forces, analysts put in a great deal of overtime, and various other measures may be taken in an attempt to cover everything. Even so, it is very difficult in such circumstances to ensure that items are not overlooked, even when their relevance or significance is readily recognized, and when their significance is obscure or debatable it is not surprising that they may be set aside. No one who has not worked with a developing crisis involving extraordinary military and political developments being reported by a geared-up collection system has any conception of the sheer volume of material which pours in. A ten-fold increase in items which the analyst should do something with, that might be important, is by no means unusual. Analysts, often close to exhaustion, may recall items of weeks back which they know are related to their current work, they may be acutely aware that there is research which should be written, but they literally cannot find the time or energy to do it. Even worse, perhaps, the analyst is so pressed that he does not have the time to contact or even learn of the existence of others who might have additional information or who might already have done some research which would assist him. When it is most needed, communication may break down for sheer lack of time.

Even more insidious may be the less obvious impending crisis where the inter-relationship of developments is not readily apparent, and particularly where two or more major geographic areas may be involved. In such cases, the difficulties of conducting needed research are greatly compounded by the fact that items from two different areas—particularly if they seem relatively obscure or questionable at the time—may not be brought together at all. Their inter-relationship is not detected until well along in the crisis, if at all, and only retrospective analyses bring together great amounts of relevant information, the significance of which was not appreciated when it was received.

The foregoing generalities are well demonstrated in two major crises of recent years, the Soviet invasion of Czechoslovakia in 1968 and the Cuban missile crisis in 1962. In the Czechoslovak situation, the intelligence community had a tremendous wealth of information, much of it of high reliability and validity, on both political and military developments which led up to the invasion. Yet a number of important pieces of information which were extremely relevant to assessing Soviet intentions apparently were never reported to the higher officials of the intelligence community, let alone the policy maker. One reason for this (although fundamental errors in judgment were probably a more important reason) was the sheer volume of material which was received. It was impossible to report everything, and in the selection process some critical indications did not make the grade.

In the Cuban missile crisis, on the other hand, the intelligence community was confronted with a series of anomalies in Soviet behavior beginning in early 1962 which raised tantalizing questions in the minds of perceptive analysts but whose relationship to Cuba was only to become apparent months later. Up almost to the time that the strategic missiles were finally detected in Cuba, two separate groups of analysts (a Soviet-Berlin group and a Latin American group) were conducting largely independent analysis, which for the most part failed to recognize that the apparent Soviet expectation of a settlement of the Berlin problem that year might in some way be related to forthcoming developments in Cuba. Thus much of the basic research which might have connected these developments was not done until after the resolution of the crisis. Only in retrospect was it apparent that the extraordinary raising of combat readiness of Soviet forces in September probably was timed with the arrival of the first strategic missiles in Cuba. And much of the research on the likely areas and possible nature of Soviet military activity in Cuba was done on a crash basis in the week between the discovery of the missiles and the President's announcement of their discovery.

It is impossible, in a brief discussion such as this, to give adequate stress to (let alone examples of) the importance of research to the warning process. There are many reasons why warning fails, or is inadequate, and it would be unfair to single out the failure by analysts to initiate and produce the requisite research as the major cause. In many cases, it may not be the analyst but rather his superior or the system itself which is primarily at fault. Nonetheless, whatever the basic causes or extenuating circumstances, insufficient research or the failure to have brought together into a meaningful and coherent pattern all the relevant intelligence must rate as a major cause of inadequate warning. The indications analyst, and others associated with the warning procedure, should never take it for granted that others know all the information available or have truly understood the facts and their implications. For the chances are that most people have not.

The greatest single justification for the existence of separate indications offices or the employment of warning analysts is that they are devoting their full time to research in depth without the distraction of having to fulfill a number of other duties. The warning analyst should never lose sight of the fact that this is his *raison d'etre.* It is difficult enough to come to a sound warning judgment when all the facts have been considered; it may be impossible without it.

b. Warning is an assessment of probabilities. As we have noted above, it is a rare instance in which impending hostilities will be so evident, or the intentions of the aggressor so unmistakable, that warning is a virtual certainty—or that everyone "has warning." It is likely that there will be some degree of uncertainty concerning the plans or intentions of the enemy even when a great amount of information is available and the collection effort has functioned extremely well. Where the amount of information is scanty, where the validity or interpretation of important data may be in question, or where there are significant delays in the receipt of material, the uncertainty normally will be considerably increased—or at worst, there may be insufficient factual data even to raise serious questions whether some aggressive action may be planned.

Although the foregoing is probably generally accepted in theory—and papers on the warning problem have repeatedly cautioned intelligence officials and policy makers alike not to expect certainty in warning—there is often a tendency to forget this important point when the live warning situation arises. Particularly because it is so important to make the right decision or right response in the face of threatened aggression, the military commander or policy level official more than ever wishes a judgment of certainty from the intelligence system—yes, the enemy is, or no, he is not planning to attack. The official on the one hand may press the intelligence system to come to a positive judgment despite the inherent uncertainties in the situation, or he may on the other demand a degree of "proof" which is absolutely unobtainable.

Now it is, of course, impossible to prove in advance that something is going to happen, when that something is dependent on the decisions and actions of people rather than the laws of nature. It could not have been "proved" in the last week of August 1939 that war was imminent in Europe, even though everyone recognized this to be so, but it could be described as a very high probability, or near certainty.

In contrast, what was the probability that the Japanese would attack Pearl Harbor, that North Korea would attack South Korea in June 1950, that the East Germans would close the Berlin sector borders (and subsequently erect the Berlin wall) in August 1961? The probability of course was actually very high; it was just that we did not know this. History shows that these occurrences in effect were considered by us to be relatively low probabilities at the time—that is, as evidenced by the lack of preparation against these contingencies, neither the intelligence system nor the policy makers considered it very likely that these would occur. We were "surprised," and, at least in the first instance, disastrously so. Yet, in retrospect, it can be demonstrated at least to the satisfaction of some that none of these events was all that improbable; they were at least good possibilities, or contingencies which should have been given more consideration than they were.

In Chapter 36 we discuss in more detail some of the factors in assessing probabilities, and in Chapter 5 there is a discussion of the related problem. of intentions versus capabilities. We will suggest here only certain guidelines for this important problem:

- In any potential warning or crisis situation, it is desirable, if not essential, that the intelligence system attempt to come to as realistic an assessment as possible of the probabilities of hostile action.
- Not only is the end assessment of value but the mere exercise of attempting to judge probabilities will bring out many useful facts, possibilities, precedents and viewpoints which might otherwise be ignored or overlooked.
- A knowledge of history, precedent and doctrine is extremely useful in assessing probabilities; and the citing of such precedents not only may bolster a case but also may tend to make the timid more willing to come to positive judgments. (If it can be stated that the military doctrine and practice of the nation in question calls for a 3-to-1 ratio of forces for attack and that it has just accomplished such a buildup, this will reinforce a judgment of probability of attack which might not otherwise be reached.)
- It is very important in reaching judgments to recognize the limits of our knowledge and collection capabilities and not to expect the impossible. Our capabilities to assess the likelihood of a Soviet move in Europe, for example, are better than against China, by virtue of better and much more expeditious collection. Our understanding of what goes on in Tibet or North Korea is minimal. Assessments of probabilities of action in these various areas must take this into account and be worded accordingly.
- Policy makers must recognize that warning cannot be issued with absolute certainty, even under the best of circumstances, but will always be an assessment of probabilities. They must realize that they will usually have to accept judgments which are less firm, or based on less hard evidence,

than they would wish, but that such types of assessments should be encouraged rather than discouraged.

c. **Warning is a judgment for the policy maker.** It is an axiom of warning that warning does not exist until it has been conveyed to the policy maker, and that he must know that he has been warned. Warning which exists only in the mind of the analyst is useless. He must find the means to convey what he believes to those who need to know it.

From the policy level, it is this factor probably more than any other which distinguishes warning intelligence from all other intelligence. The policy maker can and does get along without a vast amount of information which is compiled by the intelligence community. Some officials more than others will be receptive to intelligence information and will seek to learn details on such subjects as order of battle, weapons, internal political developments, economic plans and so forth. By and large, considering their numerous responsibilities, most policy officials are surprisingly well informed on the details of many subjects. For the most part, however, that which they are shown will be fairly carefully screened and condensed to the essentials which they most need to know, unless there is some particularly critical subject of national priority. They can function quite adequately most of the time even if they do not know much about the capabilities of the Chinese infantry division, the organization of the Soviet rear services, or the Polish airborne troops. It has been observed that a great deal of the effort of the intelligence community in recent years is essentially inhouse and consists in exchanging information with or preparing briefings for other intelligence personnel rather than higher officials.

Not so in the event of an impending crisis which may involve the security interests of the country or our allies or which could entail a commitment of US forces. It is essential that the possibility of such a development be clearly, and often repeatedly, brought to the attention of the policy official as the situation develops and that he be left in no doubt as to the potential gravity of the situation or what it might entail for national policy. In these circumstances, more rather than fewer facts, specific rather than generalized assessments, clear and realistic descriptions of the various alternatives rather than vague possibilities, and firm and unequivocal statements of the enemy's capabilities and possible or probable intentions are required.

Intelligence writers and briefers must remember that policy officials are very busy, and that assessments which carry no clear or explicit warning of what the enemy is likely to do may fail altogether to convey what the writer intended. Assessments which state that the enemy can do such-and-such, *if* he chooses or decides to do so, can convey a sense of uncertainty or even reassurance that no decisions have yet been reached, when in fact the bulk of evidence is that the enemy probably has already decided to do it. Phrases suggesting ominous possibilities which are buried in the texts of long factual discussions do not provide much warning to the policy maker who may have had time only to read the first paragraph. It is not unusual for an agency seeking to demonstrate in retrospect what a fine job of reporting it did on some subject to cull such phrases from its publications for a period of weeks. Taken in isolation or out of context, they may indeed present an ominous picture, but their impact will have been lost unless they were singled out and repeatedly emphasized at the time so that policy officials could not have failed to get the message.

A distinguished supervisor in the field of political intelligence observed many years ago that, no matter what went wrong, it was always the fault of intelligence. When disaster struck, you might remind the policy official that he had been warned of the possibility or that it had been mentioned this might happen in several briefings in the past month. And he would reply, "Well, you did not say it often enough, or loud enough."

Where warning is concerned, intelligence must be sure that it is saying it often enough and loud enough. It cannot assume that, because it issued a qualified warning last week, it is unnecessary to repeat it this week. Warning has failed more than once simply because what the analysts really thought, and were saying to one another, never was put into print. Or, if it was, it was so caveated in "coordination" or by a series of editors that what the analyst meant to convey was lost. When military action is threatened, it is no time to mince words. If the policy maker has not gotten the message, it is quite likely not his fault but that it was never made clear in the first place.

d. Warning is a conviction which results in action. We will assume now that the intelligence system has performed commendably—it has collected the data, it has done the exhaustive research required, it has come to a judgment that a military attack is probable, and it has conveyed this judgment to the policy maker in both its estimative and warning publications. And what is the purpose of all this? The purpose is to enable the policy maker to make the best possible decisions in the light of the facts and judgments sent to him, and if needed to take military and political actions to counter the threatened attack. If he is not convinced, or for some reason cannot or does not take the necessary action, the intelligence effort will have been in vain. Troops caught in an offensive which was predicted by intelligence but ignored by the commander are just as dead as if the warning had never been issued. In these circumstances, no matter how brilliant the intelligence performance, the nation will have failed if no action has been taken. For this is the ultimate function of warning.

It would be rare, although perhaps not unheard of, that intelligence might issue an unequivocal warning which would go totally unheeded by the commander or policy official. A more likely circumstance is that the warning was not explicit or clear enough (see preceding sections) or that there was a serious misjudgment of either the timing, location or nature of the attack, a matter which will be discussed in later chapters. In any case, assuming action could have been taken to avert disaster and was not, there has been a combined failure of intelligence and command or policy.

A more frequent situation in recent years has been that the policy maker—perhaps because he is dissatisfied with or distrustful of the now impersonal machinery of intelligence, or perhaps only from intellectual curiosity and because he wants to know more firsthand—makes his own intelligence. This may have its advantages, in that it is immediately responsive to policy needs, although it is clear that there could also be dangers in this. It does appear that actions have been taken at the policy level to which intelligence contributed little directly, or that policy makers have run ahead of the formal processes of intelligence in taking action to forestall possible threatened actions of enemy or potentially hostile states.

Now, not all threatened military actions are a potential threat to US security interests. Neither the Soviet suppression of the Hungarian revolt in 1956 nor the invasion of Czechoslovakia in 1968 posed any military threat to the US or NATO, and no military actions were required or even desirable on our part. Politically, however, it would have been nice to have had a little wider understanding of the likelihood of these actions, particularly the invasion of Czechoslovakia, which caused a fair amount of concern in NATO.

Regardless of how intelligence and policy function in relation to one another, or how dependent or independent the policy level may be, the important thing in the end is that appropriate action is taken, when needed, to protect the interests of national security and the security of our allies. Without this, the warning system will have failed no matter how brilliant the collection and analytical effort may have been.

Illustration of the Foregoing Principles

Some of the foregoing principles may be illustrated by the Chinese intervention in Korea in October–November 1950, an event which remains a classic example of the nature of the warning problem.*

---

*Chinese Communist forces were first contacted in Korea on about 26 October. They launched their massive offensive on 27 November.

The writer has talked with officers who were in Korea in the fateful month of November and with analysts who have researched the records of US front-line units. Their usual statement is: "Why of course we had warning. Our forces were in contact with the Chinese. We had prisoners, We had identified the units." And this is true.

The military analyst will look back at the voluminous reporting on the buildup of Chinese forces in Manchuria and on the numerous reports and indications of preparations for war throughout the Chinese mainland. He will recognize that intelligence was slow in accepting the buildup and that the total Chinese strength was underestimated, but that nonetheless a substantial buildup was accepted both in Washington and by General MacArthur's headquarters. And he will say, "Of course we had military warning."

The political analyst will recall that the Chinese summoned the Indian Ambassador in Peking on 3 October and told him that Chinese forces would enter Korea if UN forces crossed the 38th Parallel. He will look at the sharp change in the Chinese Communist and international Communist propaganda line in early November to an all-out support of the North Korean cause. And he will say, 'What more political warning could we have expected?"

The warning analyst will note that President Truman and General Mac Arthur at their meeting on Wake Island on 15 October both brought incorrect assessments—i.e., that Chinese intervention in Korea was unlikely. But the analyst also will reexamine the judgment reached on 15 November by the inter-agency intelligence committee then responsible for reaching an assessment of enemy intentions. And he will conclude that the judgment reached by that time (or 10 days before the Chinese onslaught) did provide substantial warning of the likelihood of major Chinese intervention—a warning perhaps not as clear or loud or unequivocal as it might have been but still a substantial warning. He will ask, "What happened to this? Who read it? Who believed it? Was most of the intelligence community in accord with this judgment? Were the Joint Chiefs of Staff and the President convinced?" And no one, then or now, will really be able to say who read it, how many believed it, or what weight it carried. But they will concur that "a warning" at least was issued.

Why then, is the Chinese intervention in Korea universally regarded as a great intelligence failure which contributed to a near military disaster? It is because *no action was taken* on any of this intelligence or judgments, because no orders were issued to halt the precipitate advance of US and Allied forces toward the Yalu, because no measures were initiated to prepare military defenses against even the possibility of a major Chinese onslaught. It did not matter, it must be emphasized, how much intelligence was available or who issued a warning judgment unless it resulted in such positive action. A single staff officer, if his warning had carried conviction at the right time and to the right person (in this case, to General Mac Arthur), could have significantly changed the course of events regardless of any other judgments which might have been made by the intelligence community in Washington.

4

Warning and Collection

IT WOULD HAVE BEEN POSSIBLE to add another section to the preceding chapter under "What warning is not," entitled "Warning is not collection." While true in part or in many cases, this would nonetheless have been an inadequate treatment of so vital a topic. For warning, of course, is most heavily dependent on collection, and indeed so much so that the tendency to regard them as more or less synonymous is understandable.

If collection could function ideally—with total access not only to the enemy's military dispositions and preparations but also to the political decision-making process—warning would be virtually synonymous with collection. Assuming that there were no danger of losing these collection assets, a large part of the rest of the intelligence system also could be dispensed with. If we can obtain a copy of the order of battle with details on equipment and training and supplies, most of the research effort which goes into these topics is of course unnecessary. If we could be sure of obtaining on a timely basis a transcript of cabinet or politburo meetings or copies of high-level political documents and military orders, then clearly much of the effort to deduce the nature of these from second-hand sources or agents would be superfluous. The reason why we devote so little intelligence research effort to the military forces and political activities of our allies is not just that they pose no threat to us (or at least not usually), but because we normally have fairly ready access to most of the information which we need. We may be occasionally stunned that we do not (as in the British participation in the invasion of Egypt in 1956) but ordinarily this is not a major problem.

To state the obvious, it is the effort of our enemies and potential enemies to conceal from us so much of their military activity and political decisions that makes most of the collection effort necessary in the first place. And it is because this collection is imperfect that the warning effort requires more than collection and is also a sophisticated—and inexact—analytical and reporting process.

It has been interesting to note the differing philosophical approaches to this question over a period of years. From time to time, the intelligence community has been asked to prepare analyses of our "warning capabilities"—either in the form of national estimates of our ability to provide warning of attack by the USSR or some other Communist nation, or as special studies. Some persons approaching this problem have tended to look on the question almost exclusively as a collection matter. Some studies have been addressed almost entirely to what various collection methods or systems can tell us about military or political happenings, say in Communist China, and have largely ignored the question of what the intelligence analyst and policy maker do with this information after they receive it. Such studies hopefully may give us a fair appreciation of how long it might take to find out that major troop deployments were under way toward the Taiwan Strait, or how much (or more likely, how little) access we may have to the political decision-making machinery in Peking. But they do not really come to grips with the question of what warning we might actually be likely to give that a decision had been made to take military action, or what the scope and objective of that action might be. These studies thus usually tend to convey a considerably oversimplified picture of the warning problem, and would better be described as estimates of our collection capabilities rather than our warning capabilities.

On the other hand, some persons approaching this problem have virtually ignored our collection capabilities (on the grounds in part that these may be so fluctuating and unpredictable) and have avoided coming to any kind of assessments of our warning capabilities in a specific circumstance. Such studies approach the problem almost entirely from the philosophical standpoint of the analytical problems involved, the difficulties of determining intentions, and the uncertainties which plague the warning problem from beginning to end. One group working on a major warning study on the USSR a few years ago declined, by almost unanimous vote, to address the question of how much warning the intelligence community could give the policy maker of a Soviet attack in Europe. Not only that, it declined to deal even with how much warning we could give that a major Soviet reinforcement of Eastern Europe was under way—without regard to whether or not it was intended for attack against NATO. In another high-level study, it was concluded that it would be "impossible" in the event of Warsaw Pact mobilization to determine whether the action was defensive, an attempt to coerce, or evidence of a firm decision to attack. This assessment was not made contingent in any way on what our collection capabilities might be.

Now, in the view of the experienced warning analyst, a reasonable position lies somewhere between these two extremes. Our potential ability to give warning of hostile action is very dependent on our collection capabilities against the nation or area concerned, but it will be rare indeed that our collection in and of itself will provide us unambiguous or unequivocal warning.

Even a cursory look around the world should convince almost anyone that the likelihood of our issuing timely and reasonably accurate warning should be much better in some areas than in others in large part because our collection is so much better and so much more timely. We have a much better chance of forecasting Communist action in Laos than in Tibet, or in Eastern Europe than along the border between Mongolia and Communist China. And this is almost exclusively the result of far better collection. Nonetheless, it is also true that superior collection and large quantities of high-grade information are no guarantee that firm warning will be issued, as witness the invasion of Czechoslovakia. As we have said above, warning is not a compilation of facts.

Collection Factors Most Critical for Warning

What collection assets, or qualities in collection, are most valuable for warning? What is it that the collection system can contribute which will be most useful in coming to a warning judgment? Without entering into a discussion of specific collection systems and their relative merits and drawbacks, which could be the subject of separate volumes, we may note certain general characteristics of collection which are most valuable for the analyst in a crisis or warning situation.

a. Dependability of source and evaluation. Nothing is more frustrating, misleading or time-consuming for the analyst than to receive large numbers of reports from unidentified sources of unknown reliability or unspecified access to the information they purport to know, which are received from the collector without any evaluation or comment which can assist the analyst in making any kind of assessment of the data. Regrettably, crises or potential crises tend to generate this type of reporting, for at least three different reasons. First, the collection system itself is usually geared up, so that more sources are sought whose knowledgeability or reliability is uncertain or cannot yet be determined. Secondly, the knowledge of a crisis in itself will generate a lot of "volunteer" reporting, which may range from very good to totally undependable, including some which may be the product of "paper mills" whose sole reason for existence is to fabricate reports for money. And third, the enemy's intelligence and security services for reasons of their own may be engaged in planting reports.

It is more than ever important, in such circumstances, that the collection system be evaluating as precisely as is possible the knowledgeability and dependability of its sources and transmitting some assessment of how the information was acquired and what is known about the source to the analyst. Cryptic identifications of sources by code numbers which mean nothing to the analyst, and evaluations such as F-3 with no explanation at all of where the information came from, are almost useless. Further, such insufficient sourcing or evaluation compounds the always present danger of seeming "confirmation" of information by ostensibly independent sources which are in fact the same.

At the other extreme may be the highly reliable information, from sources of proven access and dependability, which would normally be compartmented or restricted but which is released in normal channels in time of crisis because of its great value. It is lamentable but true that the collector has sometimes been known to run this risk of potential compromise of his source and yet the analyst who received the information did not know how reliable it really was, and failed to accord it the value it deserved.

In time of impending crisis, more rather than less communication is needed between the collector and the recipient, so that valuable time and energy will not be wasted in pursuing false leads, duplicate reports, or highly questionable information. Some of the restrictions which are often placed on identification of sources, particularly those not necessary for security, need to be relaxed.

Where there is reason to believe that information is being deliberately fabricated or planted, the collector will have a particular responsibility to distinguish that which is from a probable "paper mill" (this will often come through a third party), and that which appears to have originated with the enemy's intelligence services. If it can be established with reasonable certainty that someone is fabricating reports purely for monetary gain, then hopefully he will be cut out of the system. If the enemy is fabricating information for purposes of confusion or deception, on the other hand, it is very important that the analytical level recognize where the material is coming from and what its purpose may be.

Probably the greatest progress in intelligence since World War II has been in the sophistication of the collection system. And this applies not only to great technical breakthroughs in collection, such as in photography, which today provide so much valuable and irreplaceable information. The collectors also have made great progress in weeding out the junk which formerly so plagued and confused the analysts. In the summer of 1950, the system was literally flooded with fabricated or otherwise false reports concerning Chinese Communist plans and intentions, to the great detriment of the analysis of the reliable information which was available. It is likely that well over half the analysts' time that summer was spent in reading, "analyzing," and reporting this trash. Not only has the collection system long since fired most of these sources, it also now usually provides the analyst with a far better description of the source and its dependability than was true twenty years ago.

b. Value of direct access or observation. There is nothing like a live warning crisis to convince the analyst of the value of first-hand observations or access by reliable sources who are under our control and preferably who can be asked for further coverage and details. No amount of reporting from other sources—such as agents or informants, third nations no matter how reliable, the press and other open sources—can take the place of the collection method which provides for such direct access. There are three general types of sources which may provide such access or observation: military attaché and other diplomatic personnel accredited to the country in question; intercepted communications or acquisition by other means of military and political documents, decisions or orders; and photography.

The value of such sources has been universally recognized, and it should not be necessary to belabor the point. Thorough coverage of an area of known or suspected military buildup by a qualified military attaché will likely be more valuable than a dozen second-hand reports from sources of lesser competence even if they are known to be reliable. One photograph usually *is* worth a thousand words. The acquisition of valid military messages, documents and orders, etc. is absolutely invaluable. It is from this type of dependable information that the analyst can have some confidence that he is coming to a reasonable judgment of military capabilities and plans, and in some cases such collection may also provide him quite explicit insight into political attitudes and decisions.

The only reason for stressing this rather obvious point is that it is sometimes forgotten by those who should know better. Somewhat sensational reports from sources of uncertain reliability or who may have axes to grind often receive attention out of all proportion to their probable validity. Because such reports also are often short and relatively simple, they may receive rather wide electrical distribution and attract the attention of high-level officials. Meanwhile, other less "interesting" but much more useful and dependable data may receive relatively scant attention. Attaches often have been surprised, not to say chagrined, that their first-hand observations did not receive more credence back home, and that low-level reports which they considered demonstrably false had attracted so much attention.

One cardinal principle which both the collector and the analyst may do well to keep in mind is that good sources and good data reinforce one another. Provided reasonable access is available (and this is an important proviso), it will nearly always be found that a significant development, and particularly any major military preparations or movements, will be reported by more than one source and that they will be in essential if not detailed agreement. The difference in the quantity and quality of information when something really begins to happen, as opposed to the nebulous uncertainties which surround unconfirmed rumors, is often most impressive. The pieces will begin to fit together, and it is not unlikely that dozens of items of corroborative intelligence will be received—if not all at once, over a period of time. It is the measure of good collection that this is so, and it is one measure of a good analyst that he can distinguish fact or probable fact from fiction.

It is when this type of direct access or observation is not available or is limited or delayed that the greatest collection and analytical problems are likely to occur. In these circumstances, the analyst may be forced to fall back on second-hand reporting of uncertain reliability, or on information from other sources which is potentially good but inadequate. One mark of a possible impending military move of course is that the enemy will normally tighten up security measures, so that those first-hand observations by trained military observers may be denied us. Cut off from the kind of evidence he would like to have, the analyst may have to come to conclusions on less satisfactory information, and will have to qualify his judgments accordingly. It is important in this case that he also convey to higher officials the inadequacy of his facts or the deficiencies in the collection which may make him so uncertain.

It is not unlikely, even in these circumstances, however, that there will be a number of relatively low-grade reports which are mutually supporting. Although individually they may be of uncertain dependability, their sheer volume may often give them credibility. Experience has shown, for example, that persistent reports from local citizens or other sources of troop movements through a given area are likely to have some substance even if details of the size and type of forces are conflicting or exaggerated.

c. Specificity of detail. In addition to high quality first-hand access, or as a concomitant of it, the analyst confronted with a potential warning crisis will wish an abundance of specific detail. Exact counts of vehicles in convoys, careful descriptions of various items of equipment (if the observer is highly qualified, he will also usually know their exact names and models), explicit details on aircraft or naval units, precise analyses of rail traffic, these and dozens of other military details may be the determining factor in whether the analyst can make a good judgment as to the scale and type of military activity under way. Such details, always sought by order of battle analysts, may be even more essential in a period of military buildup when there may be so little time to recheck the information. In late July 1968, when the USSR began a major deployment of forces into Poland for the subsequent invasion of Czechoslovakia, a number of US and other friendly tourists observed the movements and promptly reported them to the US Embassy—but in such inadequate detail that it was really impossible to judge how much was combat troop movement as opposed to logistic movement, and how many units might actually be involved. This, of course, is not a criticism of the tourists who were not trained for this and went out of their way to report the movements at all. It is simply an illustration of how essential detail is for military analysts, and of the obligation of the collector to do everything possible to report such details in a period of military buildup.

The ability of a source to report details and the right details accurately quite often will be a measure of whether he is to be trusted at all, as well as of his powers of observation. This has long been a recognized precept for evaluating defectors. The bona fide defector from a Communist state, particularly an enlisted man, normally will have a detailed knowledge only of his own unit (but should have that), and perhaps a limited knowledge through friends or other observations of some other units. More extensive knowledge, especially of purported military plans, will normally make him suspect, unless he is a high enough ranking officer that he could reasonably be expected to have that kind of access.

This same principle may often be applicable to a covert source. The number of people in a Communist state or other dictatorship who will have access to unannounced political decisions or military plans is very small indeed. Any source who claims this sort of knowledge without some proof of such access is highly suspect, and it will quite likely be found that he also lacks knowledge of a great many details about people, places and terminology which he should know if he had that kind of access. During the Korean war, an emergency meeting was held at the request of the White House to evaluate a report from a source who purported to have knowledge of a planned Communist attack on Japan. Apart from the unlikelihood that he would have had access to such plans, it developed that he had submitted an earlier report of a purported "order" of an alleged Soviet military headquarters in Manchuria which was so patently false and inconsistent with Soviet military terminology that his credibility could be demolished forthwith.

This attention to detail by both collector and recipient of the report is vital. A crisis is a time to cut down on repetitious Sitreps relaying the same information back and forth, and to concentrate instead on providing all the available details the first time which the analyst will need to know.

d. Timeliness. The despair of the indications analyst is the information which comes in just a little too late, quite often just hours after the military action has been initiated. Sometimes this information was highly critical and, if received in time, might indeed have permitted a judgment that military action was likely or imminent. Other times, it just seems this way, since in retrospect of course the relevance of particular pieces of information becomes extremely clear. Other information which comes in late but which is wrong or inconclusive tends to be forgotten, but oh, those items that were right. Why didn't we have them sooner? Investigations of intelligence "failures" often have centered on such delays and on tardy items which may suddenly have assumed an importance far in excess of that which would have been assigned them if they had been received earlier. Like the prodigal son, this one is more important than those which were always in the fold.

Setting this human problem aside, however, there is no question that the timeliness of collection and promptness of reporting may be a decisive factor in warning, and in some cases the difference between success and failure, or between victory and defeat. This applies in particular to the crucial piece of evidence, or virtual "proof" that enemy forces are indeed embarked on a course of aggression or are preparing for imminent attack.

Now, it is not often that a single piece or act of collection, planned or fortuitous, can make this much difference, but it can happen. In such a case, the right collection, and at the right time, can literally change the course of history.

The most conspicuous example which comes to mind is Pearl Harbor. Regardless of any other failures of assessment or of planning, military or political, a single act of collection could have saved the US Fleet. This would have been a continuing reconnaissance effort, by carrier-based aircraft and from the Hawaiian Islands, to locate the Japanese Fleet, which had been missing for days. It is hard to conceive that the detection of the Japanese task force approaching Hawaii would not have led at least to the dispersion of the US Fleet from Pearl Harbor, and it quite likely would have significantly altered the course of the war.

An earlier collection "break" in the Cuban missile crisis also might have substantially altered the course of that confrontation. This has sometimes been cited as the only case in history in which a single photograph provided the ultimate proof of the enemy's plans, and which moreover was the only kind of evidence of those plans which would be totally acceptable or convincing. It is quite likely true that no amount of circumstantial evidence or seemingly reliable reports of Soviet plans could have been used as proof for presentation to the United Nations and that this photographic confirmation was essential to a good "legal" position for the imposition of a blockade, or "quarantine," by the US. But some earlier evidence that the Soviet Union was preparing to move missiles to Cuba, which in theory at least might have been obtainable, could—again in theory—have led to earlier and stronger US warnings and actions which might have forestalled the actual delivery and emplacement of the missiles. Who can say what might have happened if by some happenstance a Soviet ship en route to Cuba with strategic missiles had run aground in the Turkish Straits and its cargo revealed?

One does not have to cite such dramatic examples, however, to demonstrate that timeliness of collection may be vital for warning. This may be true both at the strategic level (in forecasting that attack is likely at all) and at the tactical level (in anticipating the likely timing and area of the attack).

The Management of Collection in Potential Crises

Several useful precepts for collectors (and recipients of their product) in potential crisis situations should have become apparent from the preceding discussion. There are other considerations as well, however, and it may therefore be useful here to point out some of the more important steps in insuring that the best possible collection effort is brought into play and that the results are promptly disseminated:

- It is important that the machinery exist for a close coordination between the collector and the analysts, to ensure that the right questions are really being sent to the field and that the collectors understand the priorities, as the analysts see them. Longstanding EEI's and routine collection requests must be set aside, if necessary, for emergency collection. Priority requests must not be delayed by excessive requirements for coordination, or "validation" of the necessity for this collection. This is the time to cut red tape, not add to it.
- As noted above, a potential crisis should call for more specific and detailed evaluations of sources, and the dispatch of this information with the information report itself, not separately. If the analysts have to send a further query about the source, which can well lead to a 48-hour or more delay in receipt of the reply, it could be too late or the two pieces of information may never be brought together.
- It is highly desirable also that the collector and the users of this information meet and have some chance to talk with one another. It may be particularly useful that the managers of covert sources be permitted to provide some better understanding (with appropriate security safeguards)

to the researchers of just what they can collect and how dependable their sources are, or are not. Such contacts can also readily clear up questions such as whether the source can or cannot be contacted for further information, and how long it might take to receive it, etc.

- Inter-agency coordination of collection requests is extremely desirable in potential crisis situations, and indeed may be indispensable. It is curious that, despite the great amount of inter-agency coordination which exists at the research and reporting levels of intelligence, there is often very little effort made to ensure that specific collection requirements are being dispatched to the field by some agency, or by the agency best qualified to obtain the data. On the one hand, a fairly simple collection requirement may never be dispatched at all. On the other, there may be needless duplication of effort, or the waste of valuable covert sources on something which could be readily obtained rather openly. It has been found very useful for inter-agency committees and working groups responsible for substantive warning also to have a voice in the collection process. In the process of reviewing indications in detail, collection gaps often become readily apparent. The various agency representatives at a watch meeting often can rapidly agree among themselves as to who should undertake to dispatch certain collection requirements. Such meetings also provide an excellent forum for indications analysts to propose specific collection requests for inter-agency consideration—often with very useful results.
- To ensure prompt and timely reporting and in sufficient detail, the most essential factor often may be that the field collector realize how crucial his information and its timeliness may be. If the intelligence system—and particularly in this case its field collectors—has not been sufficiently alerted to the danger or likelihood of some impending critical action, vital information may fail to be reported in time. The state of alert of the intelligence system itself is a very important factor in warning—both at the analytical and collection levels. (This subject is further discussed in Chapter 37.)

To Alert Collection Could Avert Disaster

For the intelligence system to come to any unusual state of alert, it must first have reason to do so—and that reason will often be that it has collected something unusual. And for the collection system to obtain that which is unusual, it must sometimes have been alerted to do something unusual. If the failure to undertake the right collection and at the right time can lead directly to disaster, so may the initiation of the right collection at the right time help avert disaster.

It follows that the single greatest contribution which the indications or warning system may be able to make in some cases is simply to ensure that all possible collection effort has been undertaken. It may be impossible to draw any firm conclusions as to the plans or intentions of a potential enemy, still less to convince one's superiors or the policy maker that any great danger lies ahead. But it may not be impossible to convince others that a stepped-up collection effort is desirable, or essential, to obtain more information on which to make a judgment. The earliest possible warning of aggressive action may not be any firm assessment—and usually is not—but rather a sense of unease conveyed to others that some preliminary actions should be taken against the contingency of a crisis. For the political policy maker, such preliminary action may be diplomatic soundings. For the military commander, it may be a precautionary alert or limited troop movements. For the intelligence system, it will usually be more collection. To have induced such preliminary action, to have initiated an extraordinary collection effort, may sometimes be all that any warning system can do.

If the intelligence system cannot guarantee warning, it should at least be sure that it has done everything possible to ensure that we will not be taken by surprise if the worst should occur. If the intelligence system really has done everything possible through collection to ensure that the policy and command level is duly alerted to the possibility of a crisis, it may well have fulfilled its function even if its analysis is somewhat less than perfect.

The North Korean attack on South Korea in June 1950 was a successfully implemented surprise attack at least in part because no extraordinary collection effort had been directed against North Korea in expectation of such action. It is probably impossible in retrospect to say how much better our warning could have been if more intensive collection (such as aerial reconnaissance) had been initiated, but it seems almost certain that we could have been better alerted. At least, with an all-out collection alert, the US military attaché in Seoul presumably would not have filed quite the same routine report as he did the day before the attack. The report reached Washington some hours after the attack, and it said in essence: "There is considerable movement of units and equipment north of the 38th Parallel. The significance of the movements cannot be judged but possibly they are in preparation for maneuvers."

5

Intentions versus Capabilities

A TIME-HONORED MILITARY PRECEPT, still quoted with some frequency, holds that intelligence should not estimate the intentions of the enemy, but only his capabilities. Sometimes this has been extended to mean that we *can* judge his capabilities but that we *cannot* judge his intentions.

The precept that intelligence properly deals only with, or should only assess, capabilities derives, of course, from the requirements of the field commander. Confronted with military forces which may attack him or which he is preparing to attack, it is essential that the commander have the most accurate possible assessment of the capabilities of enemy forces and that he prepare his defenses or plan his offense against what the enemy is capable of doing rather than attempting to guess what he might do. There is no doubt that battles and probably even wars have been lost for failure to have followed this principle and that the commander who permits his judgment of what the enemy intends to do override an assessment of what he can do is on a path to potential disaster. (The near disaster to US and Allied forces in the Chinese Communist offensive in Korea in November 1950 was directly attributable to the failure to follow this principle. The Allied campaign proceeded without regard to the capability of the Chinese to intervene and there were no defensive preparations against this contingency—since it was presumed, erroneously, that the Chinese did not intend to intervene.)

The validity of this concept, however, does not mean that intelligence at the national and strategic level should be confined to the assessment of enemy capabilities. For the fact is that intelligence at all levels, but particularly that which is prepared for the guidance of policy officials, is expected also to deal with the question of intentions. Not only the executive branch of the government, but also the congress and the public at large, believe that the function of intelligence is to ascertain what our enemies and even our friends are going to do, not only what they can do or might do. And, considering the cost of intelligence today in money and personnel and the potential consequences of misjudgment of intentions, it is hard to argue that the nation is not entitled to expect this.

It should be noted that nearly all inquiries into the presumed failures of intelligence or criticisms of its competence are addressed to why it did not provide forewarning that something was going to happen. Rarely does the policy maker or the congressional committee complain that intelligence failed to make an adequate assessment of enemy capabilities, even when this in fact may have been the case. The criticism almost invariably is: "You did not tell me this was going to happen. We were not led to expect this and were surprised." Or, "You mean for all the millions that were spent on collection, you were not able to tell us that this was likely to occur?" Protests that officials had been warned of the possibility or that the capability had been recognized are not likely to be very satisfactory in these circumstances. Like it or not, intelligence is seized with the problem of intentions. However brilliant its successes may be in other ways, its competence often will be judged in the end by the accuracy of its forecasts of what is likely to happen. And indeed this is what the warning business is all about.

Let us now examine the validity of the idea that we can judge military capabilities with a high degree of accuracy, but that intentions can never be forecast with any degree of certainty and that it is therefore unreasonable to expect the intelligence process to come to judgments of intent. How can we tell that something is going to occur before it happens? Is this not demanding the impossible? On the other hand, we can easily tell you how many troops there are deployed in this area and what they are capable of doing against the opposition.

Just how such concepts have gained prevalence or are seemingly so widely believed is a mystery. For in practice, as most experienced military analysts can testify, it may be very difficult to come to accurate assessments of the military capabilities of any nation, even when information on the area is readily obtainable. And when one is dealing with denied areas, where elementary military facts are never publicly revealed and the most rigid military security is maintained, it is often extraordinarily difficult to come to accurate estimates of such basic factors as order of battle or the strength of the armed forces. (These problems are dealt with at some length in Section IV, particularly Chapters 14 to 16.) Given enough time, and this often has been measured in years not months or weeks, intelligence has usually been able to come to fairly accurate assessments of the order of battle of most Communist nations. Total strength figures have proved extremely elusive, and the seeming certainty on such matters reflected in estimates over a period of years is of course no guarantee whatever that the estimate was accurate. And we are speaking here of situations where military strengths and dispositions have remained relatively static. Given a situation where changes are occurring, and particularly when mobilization of new units is in process or large deployments are under way in secrecy, estimates of capabilities as measured in terms of total force available may be very wide of the mark.

As a general rule, although not always, intelligence will underestimate the strength of forces, particularly ground forces, involved in a buildup and sometimes will greatly underestimate the scale of mobilization and deployments. In nearly all cases of recent years where subsequent proof was obtainable, this has been the case. The scale of the North Korean buildup for the attack on South Korea in June 1950, although recognized as formidable, was underestimated. Chinese Communist deployments to Manchuria prior to the offensive in Korea in late November 1950 were probably at least double that of most estimates at the time. The extent of North Vietnamese mobilization in 1965–66 was considerably underestimated. The Soviet response to the Hungarian revolt in late October 1956 was a precipitate deployment of units into that country which defied order of battle or numerical analysis at the time, and even today the total Soviet force employed never has been accurately established. Similarly, we do not know with any certainty how many Soviet troops were moved into Czechoslovakia in June 1968, ostensibly for Warsaw Pact exercise "Sumava," but it is highly likely that estimates are too low. Further, despite exceptionally good order of battle information on the buildup for the invasion of Czechoslovakia, several units were not identified until afterward.

Logistic capabilities, as measured in terms of ammunition, POL and other supplies immediately available for the offensive, are even more difficult to establish. In fact many logistic estimates are based not on any reliable evidence concerning the scale of the supply buildup but on an assumption that the requisite supplies will be moved forward with the units. Since the availability of supplies of course is vital to the conduct of warfare and indeed may constitute the difference between a high capability and no capability whatever, the movement of one or two items of supply may be absolutely critical. Yet it is usually unlikely that the collection system will provide much specific information on the movement of numerous important items. Few estimates have proved to be more slippery than assessing when a logistic buildup may be completed and hence when the enemy forces may actually have the capability given them on paper.

Consider further the problem of assessing the presence or absence of specific weapons, and particularly new or advanced weapons, capable of inflicting enormous damage. As everyone now knows, the so-called "missile gap" of a few years back was an intelligence gap the US was unable to make an accurate estimate of the strength of Soviet missile forces and in fact considerably overestimated Soviet capabilities in this field. It may be certain that Japanese assessments of US capabilities were revolutionized with the dropping of the first atomic bomb on Hiroshima. Did or did not the USSR have nuclear weapons in Cuba in October 1962? We think so, but we are not positive.

Add to these tangible and supposedly measurable factors the intangible factors so critical to the performance of armies and nations—such as the quality of training, leadership and morale—and it is still more apparent why estimates of capabilities may be so wrong. Nearly all Western nations at some time or another have been victims of gross misjudgment not only of the intentions but of the capabilities of other powers. In short, it is not a simple problem.

If capabilities may thus be so difficult to establish, does it follow that that ascertainment of intentions is virtually impossible? I have sometimes asked those wedded to this belief to put themselves in the position of the German High Command in the spring of 1944 as it looked across the English Channel to the enormous buildup of Allied combat power in the United Kingdom. Would their assessment have been, "Yes, there is a tremendous capability here, but can we really judge their intent? Perhaps this is only a bluff and they will launch the real invasion through southern France, or perhaps they have not yet made up their minds what they will do." Merely to pose the question reveals the fallacy of presuming that it is not possible to come to reasonable judgments of intentions. *Of course*, the Nazis could tell that the invasion was coming and that it would be made across the Channel. It is ridiculous to presume that such a buildup of military force would have been undertaken with no intention of using it or that no decisions had yet been made.

Now, the reader will say, you have picked the most conspicuous military example in the history of the world (which is probably true) and you can't make generalizations from this. In other cases it has just not been that easy or clear-cut. And of course that is right. The problem of assessing intentions rarely is this simple. But neither is it necessarily as difficult as many believe, particularly if one tries to look at it in terms of probabilities, precedents, national objectives and the options available rather than absolute either-or terms. As this treatise has already noted and will emphasize again, in warning we are dealing with probabilities, not certainties, and judgments should be made and worded accordingly. Nothing involving human behavior is absolutely certain before it occurs, and even a nation which has firmly decided on a given action may change its mind. But judgments can be made that certain courses of action are more or less probable in given circumstances, or that it appears that the enemy now plans (or intends) to do such and such. While predictions of the behavior of individuals and nations may be difficult, it is not impossible to make reasonable assessments and often with a quite high percentage of accuracy.

Although some analysts may not recognize it, the intelligence process every day is making judgments concerning the intentions of other nations, not only our enemies but many other states as well. Nor are these judgments confined to the national estimative or warning process. All analysts are making judgments all the time about the information which they receive and assessing whether or not it suggests that some nation may be getting ready to do something unusual or different from what it has been doing. Now most of the time nations are pursuing and will continue to pursue the same general courses of action which they have been following for some time, that is their attitudes and intentions will not have changed significantly and there is no requirement to be coming to continually new assessments or to be constantly reiterating what is widely recognized or accepted. While judgments of intentions of other nations thus are continually reviewed, the judgments are generally implicit rather than explicit. To cite an example, there may be no need for months or even years to reaffirm in print that it is unlikely that Communist China will attack the Nationalist-held offshore islands, i.e., that it does not intend to do so in the foreseeable future, when there are no indications that it is getting ready to do so. The intelligence judgment of intentions on this, as on many other subjects or areas, has been right and quite likely will continue to be right unless there is some discernible change in the situation. But it is essentially a negative judgment that there is nothing new that needs to be said because nothing new is going to happen.

The idea that the intelligence system either cannot or should not be making judgments of intentions usually arises only when there are sudden or major changes in a situation requiring a new judgment, and particularly a positive judgment whether or not aggressive action may be planned. Analysts who hitherto had been quite willing to make negative judgments that nothing was going to happen or that things will continue as they have will suddenly realize that they cannot make that judgment any more with confidence but also will be unwilling to come to any new positive judgments. They may thus take the position that intelligence cannot (or should not) make judgments of intentions, although they have in the past been doing just this, and will quite likely be willing to do so again in other circumstances. (The various factors which contribute to unwillingness to come to judgments in new situations are discussed in later chapters on the analytical method and the judgment process.)

Not only is it true that the intelligence process continually comes to judgments on intentions and that its errors are the exception rather than the rule, it can also be demonstrated that in some circumstances it is actually easier to reach judgments concerning intentions than it is to assess capabilities. This has often been true, for example, of the wars in both Vietnam and Laos. There is little doubt that North Vietnam for years has intended to move sufficient supplies through the Laotian Panhandle to sustain the combat capabilities of its units in South Vietnam, yet it has proved very difficult to estimate the actual through-put to the South, let alone the supplies which may reach any given unit. Similarly, it has often been possible to forecast at least several days in advance that a new Communist offensive effort was coming, i.e., intended, but the capabilities of Communist units to carry out the planned operations might be very difficult to determine and subject to considerable error. Captured documents and prisoners in both Vietnam and Laos have sometimes accurately described planned operations for a whole season; the intentions of the Communists to carry them out may be relatively clear, but other factors, most notably friendly counteractions, may be so effective that the capability to carry out the action is seriously disrupted.

It would be misleading to leave the impression that the writer believes the assessment of intentions is somehow more important than capabilities or is advocating that military precepts on assessing capabilities are outmoded or should be scrapped. Nothing could be farther from the truth. If there is one lesson to be learned from the history of warning intelligence—both its successes and its failures—it is that there is nothing more important to warning than the recognition and accurate and realistic description of capabilities. It is not only the field commander but also the policy official who first and foremost needs to understand the capabilities of the adversary. Assessments of intentions without due recognition first of the capability can be as dangerous, perhaps even more dangerous, at the national or strategic level as in the field. The greatest contribution which a military analyst can make to warning often may be to explain in clear and realistic language, for those who may not have his detailed knowledge, exactly how great a capability has been built up, how great the preponderance of force actually is, how much logistic effort has been required, or how unusual the military activity really is. Policy makers and those at lower levels as well more than once have failed to appreciate the likelihood of military action in part because no one really ever made it clear in basic English how great the capability was.

It is the history of warfare, and of warning, that the extraordinary buildup of military force (i.e., capability) is often the single most important and valid indication of intent. It is not a question of intentions versus capabilities, but of coming to logical judgments of intentions in the light of capabilities. The fact is that nations do not ordinarily undertake great and expensive buildups of combat power without the expectation or intention of using it. Large and sudden redeployments of forces, with accompanying mobilization of reserves and massive forward movements of logistic support, are usually pretty solid evidence of an intention to attack unless there is some really valid or convincing evidence to the contrary. The greater the buildup of offensive capability versus that of the adversary, the greater the deviation from normal military behavior, the more probable it usually is that military action is planned. Not certain, but probable. As someone said, "The race may not be to the swift, nor the battle to the strong, but it's still the best way to place your bets."

This principle is almost universally recognized when hostilities already are in progress. Once it has been accepted that a nation is committed to the waging of war, analysts nearly always are able to come to realistic and generally correct judgments of intentions on the basis of on extraordinary buildup of military forces in some particular area. It is when war has not yet broken out or the commitment to resort to force has not become clear or generally accepted, that there may be great reluctance to draw conclusions from the same type of military evidence which would be readily accepted in wartime. This psychological hurdle is a very serious problem in reaching warning assessments. In some circumstances, a surprising number of individuals will prove unable to reach straightforward or obvious conclusions from the massive buildup of military power and will offer a variety of other explanations, sometimes quite implausible or illogical, as to why all this combat force is being assembled.

There is, of course, one very important factor to be considered in a buildup of combat force prior to the outbreak of hostilities which differs from the wartime situation. Even if the potential aggressor has firmly decided to carry out military action if necessary to secure its objectives, it is always possible that an accommodation will be reached which will make the military operation unnecessary. There may be a negotiated political settlement, perhaps through mediation, or the other country may simply capitulate in the face of the threatened attack. Such an occurrence, however, does not invalidate a conclusion that the nation in question was determined to obtain its goals by force if needed, or that it had intended to attack unless a solution satisfactory to it was reached. Obviously, notions will not usually undertake costly and dangerous military operations if they can obtain the same objectives without them. It has been argued that, because the US did not attack the Soviet missile bases in Cuba in October 1962, no conclusions could have been drawn as to US intentions—and that the foregoing discussion concerning intentions is therefore invalid. This line of reasoning of course is entirely specious and avoids the issue. The US did not attack the bases because it did not have to—the USSR agreed to pull out the missiles. And the Soviets removed the missiles because they considered that an attack was likely—as indeed it was.

Another pitfall to be avoided is the tendency to give inadequate emphasis to the capability when its logical implications may be too alarming or unpopular. In any crisis or potential crisis situation, the military analyst should be particularly careful that he is in fact making and conveying to his superiors an accurate and adequate assessment of the military situation. He has the same responsibility as the field commander to judge capabilities first without regard to his personal predilections as to intentions. Nothing can be more dangerous than for the analyst to let his preconceptions of what is a likely course of action for the enemy influence his analysis and reporting of the military evidence. It is regrettable but true that some analysts who have rejected military action by the enemy as unlikely or as inconsistent with their preconceptions have also been known to downplay or even fail to report at all some evidence which indicated that a massive buildup of capability was in progress.

Power talks. Realistic descriptions of the buildup of military power often will convey a better sense of the likelihood of action than will a series of estimative-type judgments which fail to include the military details or reasons on which the assessment is based. To understand the capability, and to be able to view it objectively, is a prerequisite to the understanding of intent.

6

Problems of Organization and Management

FROM MODEST BEGINNINGS, there has been a proliferation in recent years of indications and watch offices, situation rooms or other alert centers whose function is to monitor developments on a 24-hour basis and alert appropriate individuals to items or events. These offices are found at both intelligence and operational levels, now up to and including the White House. While their functions vary, the great majority of them are concerned primarily with current situations on which action may be required immediately. By their very size, most such offices cannot undertake much substantive analysis, which is left for the most part to regular analysts assigned to area offices. There has been much confusion on this subject, and it is important to recognize that many offices have been somewhat erroneously dubbed indications centers and hence considered duplicative of other work, when in fact their real functions are primarily to perform night watch duty, liaison with geographic offices, alerting of appropriate analysts and higher officials, dissemination of incoming material, communications with other alert centers, and the like. There is no question of the need for offices of this type. But we are here concerned with substantive intelligence and with the organization and functioning of offices responsible for the analysis of indications in depth—in short with strategic warning.

Advantages vs. Disadvantages of Separate Indications Offices

On the face of it, there would seem to be many advantages to establishing separate offices or centers whose sole function is indications and warning intelligence. Nearly all investigators, official and unofficial, who have looked into the functioning of intelligence have seemed to be convinced that the advantages of such separate specialized indications shops outweigh the disadvantages. At the least, they are considered good insurance against that hypothetical day when the US may indeed be threatened with attack, as well as for a variety of lesser crises. There is no doubt that separate indications centers, and particularly an inter-agency center such as has now existed in Washington for over 15 years, do have a number of points in their favor, among them being:

a. Since they have no formal geographic divisions and are so small, they can readily bring together, at least in theory, all information regardless of source or area which may be pertinent to a given warning problem.
b. The analysts do not have to devote their time to a variety of other current intelligence functions and thus are able to devote full time to the assembly and analysis of information of potential indications significance.
c. The organization is extremely flexible and adaptable and can readily reallocate personnel from one area or crisis problem to another, and permit individual analysts to work on problems which involve two or more major geographic areas.
d. Such separate indications offices provide a continuity on warning problems which is unique in the community and, particularly where personnel have long experience, are almost irreplaceable sources of knowledge and expertise on warning crises and indications methodology regardless of geographic area.
e. They serve as a ready mechanism or channel for the rapid exchange of all available information among agencies on any particular indications problem or crisis, and provide the conference space and secretarial and other supporting personnel for inter-agency indications meetings and preparation of watch reports.
f. They provide one more check that significant indications material is not being overlooked, and they are free in theory to do independent indications analysis without the restraints and coordination problems so often imposed in large organizations.

Nonetheless, there are also very significant disadvantages to separate indications offices, some of which can be so serious as virtually to negate the purpose for which they were established and destroy their usefulness. Indeed, this problem has become more pronounced in recent years and, unless those whom the system is supposed to serve are alert to what may happen, they may find that the independent warning effort they thought they had does not exist at all.

The root of the problem is that intelligence and policy offices today are extremely large, with numerous diverse and overlapping functions and coordination mechanisms. As the system has expanded, formal coordinated papers have tended increasingly to replace informal papers of individual analysts or even the views of agency chiefs. Pressures have increased for uniformity of views within the intelligence system so that the policy level will hear the single, coordinated (and hence presumably best) opinion. As the demand for "coordinated intelligence" increases, so do the mechanisms for coordination, so that all sorts of estimates, studies and other papers are subjected to inter-agency reviews, often very extensive and exhaustive ones. From a system which involved almost no inter-agency coordination at the time of the outbreak of the Korean war, we have come to almost total coordination. The room for formal expression of differing opinions or even independent innovative analysis has narrowed. It has become extremely difficult, if not almost impossible, for low-level minority views to reach the policy level through *formal* channels.

Not surprisingly, these changes in the system of doing things have also affected the warning mechanism. And it has peculiarly affected the separate or independent indications office which is such a minor part of the great establishment.

What may happen in fact, and herein lies the potential danger for warning, is that an indications analysis becomes just another paper to be reviewed and coordinated in the several agencies by analysts and supervisors who in many cases will already have published analyses or opinions on the general subject if not the specific topic. With pressures for conformity and consistency with the already established agency position what they are, the chances that contradictory opinion will receive the stamp of approval and be forwarded up through the system to the policy level are slight at best. And the problem is compounded for the separate indications office in that it has so little status in the system, no representatives at the top to argue its views and is so small that many people have never heard of it and/or tend to forget that it exists.

To add to the difficulty, the small separate office will usually have no separate communications of its own (and hence cannot dispatch collection requests to the field even in a warning emergency except through channels of another agency). It will lack a printing shop and dissemination facilities of major agencies, so that its papers or analyses receive extremely restricted distribution, and very rarely if ever reach higher echelons of the government or overseas offices or commands. And analysts assigned to indications offices quite often are not invited to participate in analytical working groups even when they may be experts on the subject at hand.

Thus the system works, unless checked, to make ivory towers of indications shops. The intellectual challenge may be enormous and the opportunities for research unlimited, but the analyst assigned to the indications office may find himself for most practical purposes removed from the life of the intelligence community. It does not matter how exhaustive his research or how competent his analysis, if there is no accepted channel through which it can be published or disseminated. The effect of this may be to drive the most competent analysts out of the indications effort (very few have chosen to stay in it) and to attract conformity and mediocrity. The best analysts, when such a situation prevails, are likely to be those on short-term assignments (one or two years) who may do very excellent work but are then lost to the indications office, with a resultant loss in continuity.

The effect of this has been that most people with long experience in the matter incline to the view that the best analysts should remain in area intelligence offices rather than enter the indications system, since the opportunities to do and publish creative or useful research are likely to be so much better. The analyst, sometimes with long experience on a problem, who sees his work repeatedly rewritten or rejected not by his superiors but often by his juniors cannot help but become disillusioned.

Now, of course, the system is not supposed to operate this way, and many have been surprised to learn that it so often does. In theory—and this in fact is the reason for independent indications centers—the indications analytical effort receives a co-equal hearing with the current intelligence effort, just as the estimative process is co-equal with current intelligence. But to be equal, it must have the same opportunities, and to ensure that it does it must have the vigorous, active and continuing support of the chiefs of the intelligence agencies and the understanding of the policy level. Without this support, the indications system as any kind of independent watch-dog over national security has no status whatever. It may become the scapegoat for errors, but it can never actively perform the function for which it was intended. Possibly more than any intelligence function, the indications system in the end is dependent on the good will, understanding and personal interest of those at the top.

Selection of Personnel for the System

Outside investigators looking into the indications process sometimes have expressed surprise that the "best minds" of intelligence are not assigned to the effort, as indeed they rarely if ever are. The best minds in intelligence move on to estimates offices, supervisory positions, policy levels, military commends, or out of the system to private law practice, but they do not tarry in the warning system.

Some of the reasons for this have been set forth in the preceding paragraphs. Good minds cannot long be attracted to jobs which lack opportunity for intellectual expression, however desirable they may be in other ways. But this is not the entire problem.

A corollary difficulty is that the warning effort a good bit of the time has no requirement for the best minds, except perhaps on some long-term basic aspects of the problem. Most of the time, in the day to day course of events, the indications effort and the current intelligence effort are not very different and there will not be a great deal new for the indications analyst to do. The need for the independent indications effort on most current problems is not very great, since most current problems (even many hot crises) are not the forerunners of hostilities. The need for the independent indications effort (as we have noted in Chapter 1) is not very frequent, because it is not very often that nations are trying to conceal from us that they are getting ready for some surprise action which may threaten our national security or invoke our overseas commitments. The demand for genuine separate warning analysis thus is most erratic. A great deal of the time it may not be needed at all, but when it is needed it is desperately needed—and it is indeed a tremendous intellectual problem which warrants the detailed attention of the finest minds available.

The problem in essence is to identify early enough the existence of an impending crisis or major warning problem and to ensure that adequate intellectual effort is in fact brought to bear on it. Most agencies tend to set up special task forces to deal with crises or impending crises when they are clearly evident, but usually these task forces are set up too late for detailed early warning analysis. They are most usually established after the crisis breaks to deal with its operational aspects, which of course may be much more demanding of the time of analysts than the period of buildup for the event.

In separate indications offices, which are rarely reinforced with additional personnel to cope with crises, reallocations of the work load and concentrated effort on the critical problem (including much overtime) are the usual attempted solutions. No conscientious analyst who has ever worked through the buildup phase of a major crisis (such as the Chinese intervention in Korea or the invasion of Czechoslovakia) has felt that he did anywhere near an adequate job from a warning standpoint. The more dedicated and hardworking the analyst, the greater is his sense of failure in these cases. He is haunted by the memory of all the people he should have seen and did not, the analysis he did not get to, the papers he should have written and could not find the time for, the people he couldn't convince.

There is no simple solution to this problem, even if more resources of the community should be made available. Some modest reinforcement of indications offices in times of impending crisis does seem to be desirable *but only if truly qualified people are available.* The unqualified analyst, or one who lacks confidence in the indications method, not only is no help in such situations but may be a positive hindrance. Experience in crises has shown that the constructive substantive indications work is usually done by a handful of individuals. Quality, rather than quantity, is what is needed.

A note on the assignment of personnel, which will probably already be evident. The mere assignment of an individual to an indications shop will not automatically make him a diligent and competent warning analyst or a believer in the method. In fact, one of the greatest dangers to productive analysis can be the assignment of the wrong people. It is better to have no indications or warning offices than to assign yes-men or uninterested, timid or lazy analysts to the system. For they will not produce when the effort is needed and thus can provide a false sense of security that the work in fact has been done.

In a later section we discuss the basic principles of the indications analytical method, the qualities which make a good warning analyst, and raise the question whether analysts can be trained in indications methodology. Obviously, there is a great deal of work still to be done in this field. From the standpoint of management, it would appear extremely desirable to identify in advance those individuals who have shown an aptitude for such work, or at least have had some experience in it, and to have them earmarked for selection to indications or crisis work when the need arises. It may be possible if sufficient research is done to develop some type of aptitude tests. Even without them, however, experienced personnel and perceptive supervisors can identify those individuals with real aptitude and interest in such work. It does not have to be entirely a matter of chance whether the right people are assigned to the indications effort at the right time.

How Can Management Best Support the Warning Effort?

There are several things which management can do to improve and support the warning effort. The most important is to be sure that the voice of independent warning analysis is in fact being given an equal hearing with current or estimative analysis when there is a difference of opinion between them. To ensure this requires the active encouragement of responsible officials; often they must *ask* if there is a minority or differing view which they have not heard and request that this view be presented to them. To get the right answers, they must often initiate the questions. To be sure that the research is done, they must take an active interest in insuring that people, and the right people, are assigned to it. This may sound so elementary as not to need saying, but the fact is that warning has failed for no more reason than that supervisors or management failed to assign competent people to the effort or didn't ask the right questions. The assumption that all necessary work is automatically being done is a fallacy. In impending crises, a great deal of the work will not automatically be done. Even most of that which is most important may not be done for the simple reason that the situation is so different from normal and that the intelligence system is basically organized and staffed to take care of normal situations.

A second and extremely important responsibility of supervisory and policy levels is to ensure that all pertinent information is indeed made available to the warning system. The compartmentalization of information—both intelligence material under special clearances and policy or operational information by executive action—is a serious enough problem in normal circumstances. In time of possible impending hostile action, it could be disastrous. The gulf which separates the Joint Chiefs of Staff from the desk-level military analyst may be immense. The operational and policy information available to the National Security Council may be almost totally unknown to the working level analyst. All sorts of extremely vital but often puzzling information may become available in very restricted collection systems when in fact the enemy is getting ready to do something. From day to day it may not make much difference whether or not the analyst is privy to much of this material; he may make some minor slips for lack of it, but the security of the nation will not be at stake. But it is entirely different when the enemy is getting ready for some surprise action, for then the analyst's "need to know" may be as great if not greater than that of the policy maker.

If the gulf between the analyst and the policy maker or military commander is not bridged, both sides may be victims of fundamental and dangerous misassumptions. The policy maker assumes, often erroneously, that the intelligence level has all the information it legitimately needs—he may not really know how much has been denied to it by intermediate and zealous security officials, or he may not see *why* intelligence has a need to know. The top official often presumes, again mistakenly, that somewhere between the desk analyst with all his low-level facts and the top echelon "somebody" is putting all the intelligence and operational information into some meaningful pattern and that nothing has been overlooked. And if he thinks this, he will usually be wrong. The intelligence analyst, on the other hand, may have no real comprehension of how much is being denied to him or what is going on. Lacking this knowledge, he may set aside some extremely relevant information because he does not think it important, does not understand it, or wants to seek some further confirmation before he alerts his superior. And he also may presume erroneously, particularly if he has not had much experience, that "somewhere" all the essential integration of information is being accomplished. As a general rule, the sophisticated and perceptive analyst will have a greater and earlier recognition that something important is being withheld from him than will the policy maker realize that intelligence has not been alerted to the possibility of danger and that relevant information is therefore likely not to be reported to him. Good analysts develop an almost sixth sense that something is up that they have not been informed about, but this will not necessarily help them to conduct the right research and report the right information upward unless they are told exactly what is wanted and why. The tendency of most analysts, even very conscientious ones, will be to wait for some directions or to be told what it is they need to know.

These problems, of course, are not easily soluble, and few would suggest that everybody should know everything or that all operational or policy plans and actions should be made freely available to everyone in intelligence. One may be confident that information on highly restricted collection systems designed specifically for emergency situations will continue to be tightly controlled. Nonetheless, it is extremely important that the policy level recognize the seriousness of this problem and take steps to ensure that restrictions on dissemination are not capricious or arbitrary and that standing directives for the release of operational information to intelligence are in fact being complied with (often they are not). The inconsistencies in policy sometimes have appeared almost ludicrous. Newspapermen are repeatedly briefed "off the record" on important information and policy matters which are being denied to intelligence except at its very highest levels. All intelligence analysts learn early to read the US press for policy and operational information, and they usually become adept at perceiving which columnists are most likely to receive the most accurate information on which they can rely.

Possibly the single hottest piece of information collected since World War II was the detection of Soviet strategic missiles in Cuba. Although the information was extremely tightly held for a week, the working levels of intelligence who needed to know it (including of course the indications analysts) were immediately informed. Yet these same analysts could not obtain information, which was known to thousands of US servicemen and their families, on the deployments of US units to Florida and other US military preparations. Even months after the crisis, when post-mortem studies were attempting to determine when the USSR might first have learned from *our* military alerts and actions that we had found the missiles, an official request for information on the timing of US alerts and deployments was denied to intelligence.

Khrushchev's famous secret letter to President Kennedy on 26 October was never released to intelligence but was finally surfaced in Elie Abel's book on the crisis. The intelligence analyst who did not know that the *Pueblo* was off North Korea prior to its seizure, and most did not, was not likely to have alerted his superior to some minor anomaly in North Korean naval activity. The list of such examples is endless.

If intelligence is to do its job for the policy maker in his hour of greatest need, the latter must ensure that the intelligence system really knows what actions the policy level is planning and that some relevant information on operational plans and movements is released. In addition, steps should be taken before something happens to be sure that highly compartmented intelligence will not be bottled up when it is important to national security that the analysts know of it. Since most security regulations are extremely rigid and require the formality of specific clearances and briefings before the analyst can be introduced into the inner sanctum, it may be imperative that executive action be taken at a very high level to streamline the procedures and cut out the red tape when an impending emergency arises. As a matter of fact, security regulations nearly always are relaxed when the need to know becomes widespread. What may often happen is that the analyst and policy maker alike learn almost by accident of highly classified material and hence may not realize how tightly compartmented it normally is, or why. In such circumstances, security is much better served if those who learn anything learn more rather than less. The greatest dangers to security come from those who have had access to very classified material without knowing that it is very restricted or why.

Broadening the Horizons of the Indications System

In order to ensure that indications analysis does not become a compartmented shop, divorced from the real life of the intelligence system, it is very desirable that indications analysts be given opportunities to participate in some activities outside the daily or weekly routine. They should have at least some of the same chances to attend briefings, intelligence seminars, and schools, and even to travel, that are accorded to current and basic intelligence personnel. On the occasions when indications analysts have been invited to participate in substantive working groups or the preparation of estimates on topics related to their specialties, it has usually proved very beneficial to all concerned. After all, there is no rigid line between indications analysis and estimates, or between indications and basic intelligence. All are dealing with the same data for somewhat different purposes, and the benefits of exchange of views and mutual participation in projects are considerable.

A corollary to this, which as of this writing has yet to be tried, would be to give indications analysts a chance to publish studies or articles in the regular current and basic intelligence publications of the major agencies. The greatest single frustration to warning analysts probably is that there is no channel or accepted intelligence vehicle through which they can express their views, not even in agency "in-house" publications, so that the warning approach to problems really has no voice in the community today, either as applied to a specific current intelligence problem or as a theoretical analytical method. The most junior analysts in the fields of current and basic intelligence are given opportunities to write and publish which are denied to the most senior warning analysts.

The indications office may also benefit greatly from contacts outside the strictly intelligence field, for example with the operational and planning levels of the major departments. In the end, these are the people whom the warning system is designed to serve in the hour of crisis. It is beneficial for these functions to become acquainted with one another and to learn what each is doing. There is an affinity of interest between a number of major offices in the Department of Defense (such as the National Military Command Center, Systems Analysis, and the operational support offices of the services) and the indications effort. Both sides should take advantage of the opportunities for mutual education in each other's activities. The warning system has much to learn about what the operational level is doing and what kind of support it needs, and the operational and policy levels can profit from a better understanding of the nature of warning intelligence.

Also for these reasons, it is probably desirable to rotate intelligence personnel in the indications system rather than leave them too long in the effort. While this does, of course, result in a loss of continuity, this can be overcome in some degree by the maintenance of specialized crisis and indications files, the preparation and publication of warning and crisis studies, and perhaps the recall to the indications effort of experts when the need for augmentation becomes apparent. Perhaps a combination of rotation of personnel and the assignment of a few for extended tours would be the best solution. At least as the system has operated thus far, personnel who remain too long in the indications effort become frustrated for lack of opportunity to do research and to get it published, and so few people have actually been assigned to the effort that most analysts really have little idea how the system functions.

7

Indicator Lists

INDICATOR LISTS ARE BELIEVED to be a post-World War II development, although it is not unlikely that similar techniques had been used in the past. In about 1948, the intelligence agencies (US and British) began developing lists of actions or possible actions which might be undertaken by an enemy (specifically the Soviet Union) prior to the initiation of hostilities. From this beginning, the intelligence services of the US (and those of the UK, Canada and other nations) have gradually developed a series of indicator lists. The earlier lists, reflecting the then apparent monolithic structure of the Communist world, generally were intended to apply to the "Sino-Soviet Bloc" (as it was then called) as a whole. Little effort was made to differentiate actions which might be taken by the USSR from those which might be taken by Communist China, North Korea or the Communist Vietnamese (Viet Minh). In recognition of the differing nature of conflicts and preparations for conflicts in various areas of the world, however, numerous special lists evolved in the 1950's dealing with such topics or areas as Southeast Asia, the Taiwan Strait and Berlin. These early lists had varying degrees of coordination, formality and stature in the community, but most were prepared at the working level as tools for the analysts or by field agencies or commands, primarily for their own use.

Since the late fifties, the intelligence community has attempted to reduce this proliferation of lists and to use the resources of the community as a whole to prepare single, coordinated and formally issued lists. In recognition both of the Sino-Soviet split and the differing nature of the Soviet and Chinese Communist states and military establishments, separate lists are in use today for these areas.

Content and Preparation of Indicator Lists

The philosophy behind indicator lists is that any nation in preparation for war (either general or localized) will or may undertake certain measures (military, political and possibly economic), and that it is useful for analysts and collectors to determine in advance what these are or might be, and to identify them as specifically as possible. Because of the large variety of actions which a major nation will or may undertake prior to initiating hostilities or during periods of rising tensions, indicator lists have tended to become quite long and detailed. Thus, instead of referring simply to "mobilization of manpower," the list may have a dozen or more items identifying specific actions which may be taken during the mobilization process. While some of these specifics are intended primarily for collectors, they may also be very useful to the analysts. Because of the length of lists, however, it has sometimes been found useful to issue brief lists of selected or consolidated items as a guide to those actions deemed of most critical immediate importance, if they should occur.

In compiling indicator lists, analysts will draw on three major sources of knowledge: logic or long-time historical precedent; specific knowledge of the military doctrine or practices of the nation or nations concerned; and the lessons learned from the behavior of that nation or nations during a recent war or international crisis.

The first of these (logic or long-time precedent) is obviously essential. Regardless of what we may know about the doctrine or recent performance of any country, history tells us that all countries either must or probably will do certain types of things before they initiate hostilities. They must, at a minimum, provide the necessary combat supplies for their forces, redeploy them or at least alert them to varying degrees, and issue the order to attack. Depending on the scale and geographic area of the attack and on the likely resulting scope of the hostilities, they may undertake a large variety of additional military measures, both offensive and defensive. Most nations will probably take some measures to prepare their own populace and perhaps world opinion for their actions And, if the conflict is to be of major proportions or is likely to last for a prolonged period, they may also begin major economic reallocations before hostilities are under way.

Both analyst and collector, however, will wish more specific guides than these general precepts. For this, a knowledge of military and political doctrine and practice of the nation in question will be of invaluable assistance. Indeed, most refinements in indicator lists over the past 20 years have been based primarily on our growing knowledge and understanding of the military organization, doctrine and practices of our potential enemies, as derived from a variety of sources. As we learn more, for example, about Soviet doctrine for the employment of airborne troops in wartime or about civil defense planning for the protection of masses of the population in the event of nuclear war, we are better able to write specific indicators of what we should be looking for.

Equally valuable to the compilers of indicator lists, although usually less frequently available, will be the actual performance of a nation in a "live" warning situation. No amount of theory replaces the observance of actual performance. Preferably for the analyst, the crisis should have resulted in actual war or commitment of forces, since there will then have been no doubt that it was "for real." A crisis which abates or is adjudicated before an outbreak of hostile action is never quite as useful since there will always be some doubt how many of the observed developments were actually preparations for hostilities. The greater and/or risky the crisis and the greater the preparedness measures undertaken, the more useful for future indications purposes it is likely to be. In addition, the more recent the crisis, the more likely it is to reflect current doctrine and practice. Because of changes in Chinese Communist military forces and procedures as well as other factors, a study of the Chinese military intervention in Korea, invaluable as it is, cannot be considered an absolute guide to how the Chinese political leadership and armed forces might perform were such a situation to occur again. Similarly, an understanding of what the Soviet and Warsaw Pact forces did in preparation for the invasion of Czechoslovakia in 1968 is much more valuable today than a review of what the USSR did to suppress the Hungarian revolt in 1956. In fact, the most useful intelligence we have had since World War II on Soviet mobilization and logistic practice comes from the preparations taken for the invasion of Czechoslovakia. A number of refinements in Soviet/ Warsaw Pact indicator lists were made as a result of the Czechoslovak crisis.

It is thus evident that the carefully researched and well-prepared indicator list will contain a large variety of items, some theoretical, some well documented from recent practice or doctrine, and covering a whole spectrum of possible actions which the potential enemy may undertake. No list, however, purports to be complete or to cover every possible contingency. Each crisis or conflict, potential or actual, brings forth actions or developments which had not been anticipated on an indicator list and which in fact may be unique to the particular situation and might not occur another time.

Still more important, it must be understood that even a major crisis involving a great national mobilization and commitment of forces and many obvious political and propaganda developments may involve only a relatively small fraction of the actions set forth on a comprehensive indicator list. And, of these actions, the best intelligence collection system can reasonably be expected to observe a still smaller fraction. Some of the most important developments, particularly those immediately preceding the initiation of hostilities, are unlikely under the best of circumstances to be detected, or at least to be detected in time. A judgment that a specific indicator is unlikely to be detected, however, should never preclude its inclusion on an indicator list, since there is always a chance that the collection system, perhaps fortuitously, will learn of it. As someone once said, an indicator list is a desideratum, not an absolute guide to what may occur, still less a statement of what we are likely to know.

At one time, it was considered desirable to divide indicator lists, not only into various subject categories (military, political, economic) and sub-categories of these, but also into time phases—usually long-range, intermediate-range and short-range. This procedure rested on the hypothesis, which is not invalid in theory, that some preparations for hostilities are of a much longer term nature than others and require much longer to implement, whereas others could reasonably be expected to occur very shortly before the initiation of the attack. Experience, however, has shown that such pre-judgments of likely time phases of actions are of doubtful practical use, and can be misleading or even dangerous. Some preparations may be so long term (the start of a program to build large numbers of new submarines, for example) that their relevance to the initiation of hostilities at some date in the future is questionable, at best. Other time-consuming logistic preparations (such as construction of new roads which would be useful for support of military operations) may be of a long-term contingency nature or can be indicative of a relatively short-term intention to make use of them as soon as possible.

Similarly, dependent on circumstances, an action which would usually be considered very short-term in nature may occur weeks or even months before the military attack or action is finally carried out. (Some of the Soviet units which finally invaded Czechoslovakia on 20–21 August 1968 were mobilized and deployed to the border on 7–8 May and were held in readiness for operations for over three months. According to a defector from one of the units involved, troops in his regiment were actually issued the basic load of ammunition on 9 May, but it was then withdrawn.) On the other hand, in a rapidly developing crisis calling for immediate military action, all preparations may be telescoped into a few days—as in the Soviet suppression of the Hungarian revolt in 1956. Still another factor to be considered is that the collection of information on a given action may run days or weeks after the occurrence, so that actions which may appear to be recent have actually been in progress for some time.

Thus, for these and other reasons, most indicator lists today have dropped a distinction between long and short-term preparations for hostilities. At the same time, some preparations judged to be so long term in nature that they are of doubtful validity as indicators have been dropped altogether. The focus today is on the collection and evaluation of all indicators, regardless of timing, which may point to a likelihood or possibility that a nation has begun preparations for the initiation of hostilities. The assessment of the timing of these actions, individually and in relation to one another, has become a part of the evaluation and interpretation process, rather than a part of the indicator lists themselves. The problems involved in assessment and interpretation are discussed in detail in later sections of this study.

Uses of Indicator Lists

Any analyst who has participated in the preparation of a detailed indicator list and its coordination with other agencies will have learned a great deal in the process. In fact, the most useful aspect of these lists may be that the analysts who work on them must examine in detail the steps which a prospective enemy is most likely to take before initiating hostilities. For beginners, such experience is invaluable, and even analysts with long experience always learn something new in the process of preparing or revising such lists. The process serves as an extremely useful medium for an exchange of expertise between basic analysts (with their invaluable knowledge of the nuts and bolts) and current indications analysts who may be very knowledgeable on the theory of warning and the general types of things to be looking for but rather ignorant on the specifics. Thus, if nothing further were done with indicator lists, the time spent in preparing them would probably not have been wasted.

The usual reason for preparing such lists, however, is to give them wide dissemination to other analysts, supervisors, collectors and field agencies in order to guide them on what it is we want to know. At least in theory, every attaché is armed with an indicator list (together of course with all his other guidance and directives on what to look for), and each current analyst frequently consults his list to see how many indicators are showing up "positive" or where his collection gaps may be. It is not uncommon for those new to the intelligence process to expect the indicator list to be a kind of bible or at least the master guide on what to be watching for. "Let me see your indicator files" is not an unusual query.

Of course, it is not all this simple, as was probably evident from the preceding section. To attaches, the indicator list (even if they have received it which sometimes they have not) is one more piece of collection guidance and one more document to keep in the safe. Practically never will the attaché receive a query specifically pegged to an item on the indicator list. He may receive a query on a subject on the indicator list, and quite likely will, but it will not be related to the list and there will be no reason to consult it. Further the attaché or other field collector who takes time to examine the list in detail quite likely will find: (1) there are an enormous number of items on the list which he could never collect against or which are entirely beyond his field; (2) on those items on which he is competent to collect, the list is far too general to be of use to him; (3) he has a far better sense of what is potentially significant or abnormal or of indications significance in his area than could possibly be conveyed by any indicator list; and (4) when he does have something of warning significance, or potential warning significance, to report, it cannot readily be fitted into a particular category on the list, and needs a fuller explanation of its possible meaning. And of course he will be right on all counts. What the field collector needs when something begins to happen which has potentially ominous connotations is not to sit down and reread his indicator list, but rather to have specific guidance on exactly what to look for in his area (and where) and to be relieved of a lot of routine requests when he should be concentrating on what is of immediate importance. He needs to know that there may be an impending crisis (sometimes he will not know this unless his home office tells him), that he should be devoting greater than normal attention to field collection and to prompt reporting, and he needs to have lateral reporting from other areas on subjects which he may assist in evaluating.

Does it follow that an indicator list is really of very little value for small field offices and would better never be sent to them? Many might say yes, and that the proper time and place to study general indicator lists is prior to departure. This is probably a question best answered by those who have been in the field collection program. The difficulty of course is that the value of such lists can really only be judged when there is a crisis, when there are some real indications to report.

In the interests of more specific guidance and of not burdening the field office with a detailed list, some attempts have been made to compile very specific lists of things to look for in specific cities or areas in event of possible preparation for major hostilities. Obviously, such lists can only be prepared by those with the most detailed knowledge of the area. Much more remains to be done in this field, but it is one which might prove to be quite fruitful. By reducing the general to the specific and for a specific area, indicator lists in the future may prove to have greater usefulness for the field than has heretofore been the case.

And what use is the indicator list to the analyst back at headquarters? What is he doing with it? The chances are that he also has it somewhere in the office but has not looked at it very often. If he has helped to prepare it in the first place, he will probably have very little need to consult it since he will know almost automatically that a given report does or does not fit some category on the indicator list. Hopefully, he will from time to time consult his list, particularly if he begins to note a number of developments which could indicate an impending outbreak of conflict or some other crisis. In this event, even the very experienced analyst may find it helpful to compare what indications he has against those which either have not occurred for sure and those on which he does not have enough information to know one way or the other. On the latter (the gaps) perhaps some specific field guidance may be in order—can it be ascertained whether or not the military has requisitioned trucks from the civilian economy or is it just impossible to find out in this particular area? If he does not expect too much, and remembers that at best he may hope to see only a fraction of all the indicators on the list, he may find that the list is a useful guide to give him perspective on his present crisis.

In normal times, when the situation is reasonably quiet and no disturbing military or political developments are evident, an indicator list is not going to be much use—unless one needs to be reassured periodically that things are pretty normal. An indicator list is really not for all the time—it is a sometime thing. And even when that "some time" occurs—that twice a decade perhaps when we are confronted with a real warning crisis—no one should look on the indicator list as a solution to the warning problem. The mere ticking off of items on an indicator list has never produced warning—and it never will. It is a tool but not a panacea.

In the next section we discuss some other tools of the warning analyst and offer some helpful hints on what and how to file. We would offer one "don't" here. *Don't* set up a list of detailed file headings to conform with all the items on the indicator list and wait for items to come in to file under them. You will end up with lots of headings and very little to put under them, and you will have lots of reports and items which you have no place to file. Headings to conform to many of the broad general categories of the indicator list, on the other hand, will be very useful and permit one to file information on general trends or basic data on these subjects (e.g. mobilization procedures) which may be very helpful background to evaluate an abnormality when and if it appears. But a file set up to take care of nothing but abnormalities (some rarely noted and some never) will likely become a white elephant. In a lifetime career of watching indications, you might never receive a report in time that a company on the front line had been issued tactical maps of enemy territory immediately across the frontier. But if you get it, you won't need a heading in your card file to tell you that it is important. And you also probably won't have time to file it before the war starts.

8

The Compiling of Indications

THIS CHAPTER IS A DISCUSSION for basic indications and current analysts, and those not concerned with such working level details may wish to skip it. It is a response to many questions which the writer has been asked over a period of years on what should be filed and what are the best and most reliable methods of compiling indications so that data may be recovered and analyzed.

As all analysts know, it is possible to collect an enormous amount of paper in an extremely short time on almost any current intelligence problem, and the greater the current interest in the subject, the greater will be the amount of verbiage. In no time at all, the analyst can collect drawers full of raw material plus current intelligence items and special studies on an ongoing situation like the war in Vietnam. The measure of the importance of a crisis is sometimes the amount of reporting it engenders. The Cuban missile crisis or the Soviet/Warsaw Pact military buildup for the invasion of Czechoslovakia will result in an avalanche of reporting, not all of it original unfortunately. Sitreps (situation reports) repeating raw data received 48 hours before pour in on the analyst who is hard put to it to read everything, let alone put it into some kind of filing system where he can recover what he needs to know.

There are no panaceas to this problem, at least none have been found yet, and the writer does not wish to imply that the following suggestions will necessarily be the best method in a crisis or even from day to day when things are relatively quiet. Nothing yet has been found to replace a retentive memory, a recognition of what is important and what is not, and a sense of how things fit together.

There are perhaps four basic filing or compiling problems which the indications analyst should be prepared to deal with. They are: (1) the extracting for files of current raw data or information of potential indications significance; (2) compiling the highlights of such data into a readily usable form by topic or area; (3) coping with the sudden but short-term crisis; and (4) long-term warning or indications files.

Extracting Indications Data

The first principle is that the indications analyst must have the basic raw data, extracted verbatim or competently summarized if the item is very long, with the original source notations and evaluation and the date of the information (not just when it was acquired). He can, if necessary, dispense with all current intelligence summaries of such data, all special sitreps (unless they contain new data), and a variety of other reports, but he cannot do without the basic information.

An excellent system, which has proved very useful, is to assign to duty officers, with clerical assistance, a responsibility for extracting raw information deemed to be of indications value in a daily informal publication for the use of analysts and also supervisors who want a quick review of indications highlights. In addition to the items selected for inclusion by the duty officers on an around-the-clock basis, the substantive indications analysts may ask to have additional items put in, sometimes with their comments or interpretations. The individual items can then be clipped by the analysts for their own use and assembled under appropriate subject headings. An ideal method is to put them onto 5x8 cards or cut them to card size. Alternatively, they may be assembled in folders by topic, although this is usually less satisfactory. Obviously, discrimination is needed to retain what is of some lasting indications interest and to pitch out a lot of current information which may be of no interest next week.

If he lacks this type of round-the-clock assistance, the analyst hopefully should have some clerical help in extracting the data which he needs to retain.

For indications purposes, the material should be filed under the date on which the development occurred, not when it was reported or received. You are interested in the inter-relationship of events in time, not when the information became available.

As noted in the preceding chapter, the general headings of indicator lists may be good headings for some portions of the file, but specific items from the indicator list usually should not be used as file headings. Above all, however, the analyst should maintain a flexible system, in which he can start new headings or revise their titles as required, or break some category down into sub-headings. Don't get trapped in a rigid system which cannot be readily expanded or modified as new developments occur. For this reason, pencil headings at the tops of cards, rather than ink or typewriter, are preferable, and numbered headings and sub-headings should be avoided. The system should be designed to serve you, not you the system.

If you are maintaining files on more than one area, set up separate sections for each. Do not attempt to file Chinese troop movements toward the Soviet border together with Chinese troops in North Vietnam. It won't work.

The microfilming and indexing of such files by country and key words has also proved useful in recovering specific items. The analyst should remember, however, that when he keeps his own subject files he is doing much of his research as he goes along. If he relies on any library system such as microfilms, he will have to recover each item separately and really do the research from scratch. Where large numbers of items may be involved, this can be very time-consuming. The microfilm system is really of greater assistance for a spot check of some specific question, rather than as a continuing research aid.

Compiling Highlights: The Indications Chronology

Even a good card file or similar system, hopefully weeded out from time to time, will leave the analyst with a great deal of material to cull through if he wishes a succinct summary of the highlights of any developing indications problem. Further, the possible relationship of items under different headings may be overlooked and important items forgotten. Also, of course, no one else can use the material without going through all the cards.

The writer believes that no indications methodology yet devised is as useful or meaningful as the properly prepared indications chronology. The method is applicable both to relatively normal situations, for recording the highlights, and to budding crisis situations in which large volumes of material are being received. While its advantages are numerous, it must also be observed that it is a very time-consuming task requiring the most conscientious effort, which is probably the chief reason that so few are prepared.

The purpose of the indications chronology is to record briefly in time sequence (by date of occurrence, not date of reporting) all known developments, reported facts, alleged actions or plans, rumors, or anything else which might be an indication of impending aggressive action or other abnormal activity. For the initial chronology, it is not necessary to prove any inter-relationship of the various items, but merely to note them (obviously, there must be some possible connection, even if remote, to include an item). Some notation as to the validity of the information is desirable (particularly unconfirmed or dubious items), and it is often helpful to note the source as well, but the items should be as brief as possible. The chronology should also include significant actions by our side (or other nations) which might cause some reaction or help explain some enemy activity.

In a slow-building crisis of many months (which is the usual, not the unusual crisis), the inter-relationship of old and new material may become immediately evident once the information is approached chronologically. Decision times may be isolated, or the likely reason for some action may become apparent. Where serious military preparations are under way, it will quite likely appear that a number of actions were initiated at about the same time. Events which appeared to have no relationship at the time they were reported may suddenly assume a meaningful pattern. Moreover, the method provides an almost foolproof way that the analyst will not overlook something, perhaps because he did not understand it when received or did not know where to file it. A chronology is a catch-all for anomalies—and while not all anomalies lead to crises, all crises are made up of anomalies.

In addition to insuring that research is done and items not ignored, the draft chronology can rapidly be edited in final form for distribution on a crash basis, and the analyst also can readily produce a written analysis or summary of the evidence from the chronology. Where a great deal of material is coming in, a page for each day can be maintained and new items added as received, and only those pages need be rerun to ensure that it is current.

The method is also very handy for just keeping an historical record of major developments in some area or country (say the USSR and Eastern Europe, the Sino-Soviet border, or Laos) even when no particular crisis is evident.

Further, once an item has been briefly recorded, a large amount of paper can be thrown away. It is a tremendous saver of file space.

Good chronologies of major crises can be kept indefinitely, most of the other data can be destroyed, and the document will remain the most useful document on the crisis which you could have.

The Sudden Short-Term Crisis

Desirable as the preceding methods may be, the pressure of time during a major crisis will probably preclude the preparing of chronologies and possibly the maintenance of any types of files until the crisis is resolved. Analysts are likely to be so overworked simply keeping up with incoming material and current reporting requirements that the best that can be done is to sort material into some makeshift system which will permit the location of the highlights. Unfortunately, when research may be most essential and the detailed chronology would be most desirable, there is no time for it.

Put first things first. Your files are not more important than getting out analyses and summaries or comments on the significance of items which others may have overlooked. Find space to sort and keep the material you really need, and get some clerical help in doing it if possible. Call on basic analysts for help if you can get it. Let your files and even a chronology go if necessary to ensure that you are not overlooking new material which should be reported and analyzed.

Long-Term Warning Files

Unlike current intelligence files, which rapidly become out-of-date, good warning files improve with age. Some of the most valuable files in intelligence today are the few extant on the Chinese Communist intervention in Korea, since most analysts' research files of this have been destroyed and anyone seeking to review the basic data from the original raw material would have an almost insurmountable task.

One of the most useful things which an indications office can do is to keep files of crisis situations and warning problems. This may include extracts of the original material in card form as described above, and also the chronologies and special studies. Post-mortems of what happened and how it was analyzed and reported at the time are extremely useful for future study of indications and warning methodology. If such files are well compiled, virtually all other current information on the problem can usually be destroyed. Don't throw out crisis files and studies just because they are old and there has been no demand for them in years. There was little demand for ten years for studies on the Soviet reaction to the Hungarian revolt of 1956, but interest perked up noticeably in the summer of 1968. Similarly, studies of Soviet reaction in the Suez crisis of 1956 were suddenly in demand at the time of the six-day war in June 1967.

Another type of file which the indications analyst or office should maintain is the basic data on how a nation goes to war: alert and combat readiness procedures, mobilization laws, studies of major war games or exercises, civil defense doctrine and practice, and a host of other similar material which is rarely needed but of absolutely vital importance when there is a threat of employment of military force. (See next chapter, "Can Computers Help?")

9

Can Computers Help?

THE WRITER MAKES NO CLAIM to any expertise in the field of computers or automatic data processing, either with regard to the technical aspects or the types of data which should be programmed. Certain observations concerning the potential usefulness of automatic data processing for warning, however, may be made from the standpoint of the indications analyst who at least is qualified to say what it is he may most need to know in a hurry in a live crisis situation. The question of whether the programming of such data is feasible may then be left to the experts.

Generally, of course, computers are most useful and applicable to problems which either are primarily quantitative or mathematical in nature (i.e., where an exact answer can be obtained) or repetitive (where there is a broad data base against which normalcy or variations from normalcy can be measured in terms of frequency or probabilities). Computers generally are least useful or not useful at all where the problem is abstract, not measurable, unique or rarely repetitive, or involves a complex interplay of factors from which a judgment of human behavior must be made. As is also well known, an error in programming where one piece of critical information is incorrectly entered or omitted may lead to a large margin of error by the computer—as witness the mistakes which have sometimes been made on election night because the data put into the computer were not quite the right selection.

Because of these factors, most indications experts believe and computer experts have generally concurred that the usefulness of computers for strategic warning is limited and in fact could be quite dangerous if too much weight were to be ascribed to the computer's "judgment." For the warning process is not only a complex analytical problem, it is also essentially non-repetitive, non-mathematical, does involve a judgment of probabilities of human behavior, and must work from a data base which is inexact, certain to be incomplete, and in which the validity of some of the most important available information cannot be established. Further, it must be noted that the question of what data should be fed into the computer in a live warning situation would be absolutely critical to the judgment it might reach. Just as indications can be selected to support completely diverse opinions, so might the computer come to completely different judgments dependent on whether certain particular pieces of information or indications were or were not selected for inclusion. Thus the judgment of what to put into the computer might and probably would be the most critical decision of all, and the person who does that (not normally a very high-ranking or experienced expert) could theoretically have more influence on the final decision than the policy maker.

Since the computer's judgment essentially is quantitative rather than qualitative, there is also reason to suspect that it might be more vulnerable to a clever deception effort than would the alert and sophisticated analyst. The successful deception effort, of course, builds up a case which is inherently probable and believable—seemingly more probable and believable (or at least so the designer hopes) than the real plan. The analyst who sees through the deception effort, if he does, must rely on something other than quantity and superficial plausibility or consistency of the evidence to come to the conclusion that it is suspect, or that something is amiss. This sophisticated sense of perception, almost intuition, of the experienced analyst is not measurable and almost certainly never will be computerized, whatever other progress there may be in the state of the art.

Thus, the question whether a computer even under the most ideal warning conditions could be expected to do better than, or indeed as well as, the human mind or a number of human minds together is at best debatable. And whether its judgment would or should be trusted, particularly if the security of the nation were at stake, is even more crucial.

Does this mean that there is no use for computers or automatic processing of information in the warning system? On the contrary, there are a variety of ways in which the computer probably could contribute to the warning field—provided its limitations are understood and the responsibilities of coming to human judgments are not evaded.

First, it is important to recognize that warning or indications analysts are dealing with the same basic information which current, basic and estimates personnel are dealing with, and that their requirements for the processing of much information are essentially the same as well. There is nothing unique about the warning analyst's requirement for information on an airfield, a ground force unit, the defense budget, the location of SAM sites, and a host of other subjects. He will need much the same information to do his job which the current order of battle analyst or logistics expert will need. It follows that any system which will simplify or speed up the recovery of many types of basic information for the current analyst may also incidentally serve the warning analyst.

There is one important factor to be remembered in computer processing of data, however, which could be peculiarly important for warning. This is the question of timeliness and of insuring that all the new data which could be of indications value is in fact being put into the system. Experience in many crises has demonstrated that analysts and the intelligence system in general are often barely able to keep up with the flow of material for essential current reporting, let alone take care of less immediate requirements such as the filing or computerization of data. It would be optimistic indeed to presume that this problem is likely to be soluble by any administrative measures or that more people to prepare the computer data can be found in a hurry during an impending crisis.

Collection systems which have been designed specifically for tactical warning or to detect unusual deployments or preparedness measures of enemy forces in many cases collect data which is subject to computerization. Increasingly, data collected by various sensor devices and such material as electronic emissions are being computerized or at least being considered for such programs. In the period immediately prior to surprise attack or while major covert deployments of forces and other preparedness measures were in progress, there might be a substantial increase in information of this type. Obviously some central effort, assisted by ADP, to bring all such data together and to evaluate them rapidly, almost instantaneously, could be of crucial importance. Planning is essential for the day when all sorts of abnormalities might suddenly appear and when there would be no time to devise new means of coping with, evaluating or displaying the material. In the hour of crisis, it may be absolutely essential that the data processing system be as rapid as possible and that the collation of various types of measurable data be centralized. At such a time, the strategic and tactical warning problems tend to merge. Warning analysis, if it has not already done so, passes from the intelligence system to the policy maker and the military commander, if for no other reason than that time does not permit the normal process to function. Decisions must be made immediately. The potential usefulness of computers in these circumstances is obvious and it has been recognized.

Far less attention has been given to another potential use of computers to assist the analyst and policy maker in recognizing and assessing the importance of certain information in crisis situations. Because the live warning situation is unique and involves preparedness measures rarely if ever observed in normal circumstances, the community may suddenly be confronted with a number of developments and reports whose significance is not appreciated simply because they have never been seen before. In many cases, these are the true indications, the developments which distinguish *real* preparations for hostilities from exercises or which may mean that forces have been brought to an exceptionally high state of readiness, have been issued weapons which are never issued in peacetime, or have converted from a peacetime to a wartime organizational status or nomenclature.

All indicator lists and collection systems recognize this type of truly unique or exceptional preparation as extremely vital to warning. Many assets, in men and money, have been expended in an effort to acquire the enemy's classified military documents or other reliable evidence concerning such vital facts as who will control the release of nuclear weapons, where the alternate military command posts are, what the wartime designation and status of an administrative headquarters is to be, how civil defense evacuation orders will be issued, where the chemical warfare stocks are located, and what happens when a unit is brought to "full combat readiness."

Yet, when the crisis occurs, all too often this critical basic intelligence, which is so vital to the interpretation of current reporting, is not readily available. It may have been reported only once or twice before, perhaps by a defector or in an obscure document long since relegated to the library or half-forgotten in some analyst's safe. To learn the significance of some seemingly simple fact, or the meaning or validity of a military term, may literally require a basic research effort for which the intelligence system and the policy maker do not have time to wait. When the Cuban missile crisis broke, and the first reports were received that some Soviet units were at "combat readiness number one," it is literally true that those responsible for US warning judgments were unable to find out through readily available channels whether this was the highest or lowest state of Soviet military readiness. Yet the information was available in Soviet documents and was brought to light—after the crisis was over.

This is not an isolated example; it is typical of what occurs in a crisis. It may be the first and only time in history that anyone has needed to know in a hurry whether reservists are normally called up suddenly at night or notified ahead by mail, what is the exact term in the enemy's language for a certain type of chemical shell, or whether the medical support unit of a division is to be raised from company to battalion strength in the event of mobilization.

The intelligence system has enough to do in a crisis situation that it should not also have to take time out to locate such essential basic information, or even to find out whether the answer is available or not. Every effort should be made to ensure that such vital basic material is readily available, can be located immediately, is not dependent on the memory or files of some analyst who may happen to be on leave, and that the information is the latest authoritative data regardless of classification. The last point may be very important, since some data of this type may be extremely classified and thus will not be brought forth in response to an inquiry through normal channels.

It would seem that computers, while not a panacea for this problem, might do much to cut down the time-consuming effort to locate essential basic data when extraordinary developments, particularly military developments, begin to occur. Simply stated, what the computer needs to have built in is *the language of war*. The computer which is to serve warning intelligence should be programmed not to handle the routine but the exceptional, to serve as a repository for the known facts and exact terminology about how the enemy will go to war. Examples, by no means intended to be exhaustive, of the types of information which the computer might store in its memory bank are:

- A glossary of the enemy's military terms (in the native tongue and in translation uniformly agreed on in the community) to include particularly terms for various types of alerts, stages of combat readiness, mobilization, nuclear or other special weapons.
- Lists of inactive or demobilized units, to include cover names and honorifics.
- Slang terms used by enemy troops for weapons, mobilization and the like.
- Terminology of any special wartime legislation; exact names of government or military organs or commands which may be activated only in wartime, together with description of their functions; wartime designations of administrative headquarters.
- Civil defense and medical terms.
- Locations of alternate or wartime military command posts or headquarters, alternate or auxiliary airfields, naval dispersal bases or areas, assembly areas for ground force units, civil defense headquarters, governmental evacuation sites, nuclear and chemical weapons storage areas.
- Possibly, political terminology or propaganda terms either used in past crises or wars, or which research has established as indicative of rising tensions.

It cannot be overemphasized that any information of the foregoing types which is computerized must be the result of the most exhaustive and careful research by the most qualified analysts, and that the community at its highest level must ensure that compartmented data is considered as well as that which is readily available. If highly classified information serves to confirm that which is generally known, that is fine and there is no problem. If it can contribute more or if it contradicts information generally available, then it is essential that this be flagged in some way. Further, if such a computer effort is to be of use when it is needed, it must be periodically reviewed and updated by real experts in the field. An inadequate or inaccurate effort could be worse than nothing. This is no job for amateurs.

To ensure the accuracy and quality of the data n any such computer system, its contents should probably be reviewed annually under the auspices and direction of the highest levels of the intelligence system. It should be given the same care and attention, and enjoy the same degree of authority, as a national estimate or the NIPP.

10

Some Fundamentals of Indications Analysis

MOST OF THE REMAINDER of this handbook is devoted, in one way or another, to some aspect of indications analysis. We will be considering both general and specific aspects of the analytical problem, both theory and precepts and how to tackle specific problems such as order of battle or logistics in an abnormal or crisis situation. And we will also be considering the end product of the analytical effort—the reaching of the assessment or warning judgment and its transmission to the policy maker.

Warning involves several stages or levels of activity within the intelligence community. In practice, these will overlap in some degree, but for descriptive purposes we may define them as follows:

a. The collection of the raw data—obviously a necessity.
b. The selection from the mass of incoming information of those pieces which appear to have some relevance to the enemy's course of action or intentions—the isolation of the indications or the potential or possible indications.
c. The evaluation of these indications, or assessment of their validity, meaning and significance, in so far as they relate to the establishment of the relevant *facts* pertinent to the problem (for example, do these reports and indications collectively provide evidence that mobilization is in progress?)
d. The assessment of the meaning of the relevant "facts" for the enemy's possible course of action—i.e., what does the totality of the available evidence indicate, or suggest, that he is likely to do?

e. The preparation of a paper or briefing for the policy maker or military commander conveying the assessment of the evidence—i.e., the issuance of the warning judgment.

Each of these steps may be critical, and normally no one of them can be ignored without some detriment to the warning judgment. While it is possible in rare cases to move directly from the collection to the judgment of intent, it is much more usual that the final warning judgment will be a product only of the most painstaking collation of indications, and evaluation and assessment of the factual information. Normally, before we can come to a judgment of what the facts mean, we must have been through a lengthy process of determining what the facts are. To slight any step in the process can be dangerous.

It may be observed that there is seemingly very little difference between this process of analysis and judgment and that in any other field of intelligence. Why study indications analysis? What's different about it?

Many would maintain that there is no difference and that indications analysis is a fancy name for what intelligence is doing all the time. Hence, anyone can do indications analysis, or at least any good analyst can.

Interestingly enough, the warning specialist would not differ greatly with this view. As a general rule, an exceptionally good analyst in another field will also make an excellent indications analyst. A good analytical mind is usually capable of doing more than one type of analysis well. Diligence, an aptitude for detailed research, imagination and objectivity are talents applicable to any number of fields in intelligence, as well as to warning.

The difference between warning analysis and other intelligence analysis is largely one of degree or, one might say, of intensity. The analytical factors and techniques most useful or essential for warning also are required in varying degrees in other fields of intelligence analysis. Nonetheless, the analytical problems of warning, if not unique individually, pose a complexity of difficulties in combination which are certainly exceptional and would seem to warrant some special consideration.

Let us examine first some of the fundamentals of indications research which analysts, their supervisors and higher level officials as well should understand.

Recognition of the Inadequacy of Our Knowledge

Warning analysis must begin with a realistic understanding of how much—or more accurately, how little—we know about what is going on in the areas of the world controlled by our enemies or potential enemies *on a current day-*

to-day basis. Large numbers of people, in fact probably most people who are not actively engaged in collection or research on these areas, often have quite distorted ideas about what we know about the situation right now or what our current collection capabilities really are.

For the most part, people with a superficial rather than a detailed knowledge will tend to believe that we know more about the current situation than we really do. This tendency to exaggerate our current knowledge or collection capabilities is the product of several factors. Most important perhaps is that our overall, long-term, basic intelligence on our adversaries is often quite good, or even excellent. We may really know a good deal about the Chinese transportation system, the locations and strengths of Soviet units, military production in Poland and innumerable other subjects all pertinent in some degree to the capabilities and intentions of the nations concerned. What the inexperienced observer may not realize is how long it has taken to obtain such information, or that our knowledge is the product of literally years of painstaking collection and meticulous analysis. He may not understand that our best information (for example that from a high-class defector or clandestinely acquired documents) may be months or even years old before we obtain it, and that it is rare that information of such quality is available on a current basis.

It cannot be denied also that the managers of certain collection systems may tend to create the impression that they really know far more—or have reported far more—on a current situation than is actually the case. Post mortems and retrospective analyses are notorious vehicles for setting forth all sorts of detailed military facts and interpretations as if all this had really been available and in the hands of the intelligence users at the time the event occurred. Such self-serving reporting may be useful for budgetary purposes (indeed, this may be the most frequent reason for this type of distorted reporting) but it does a real disservice to the rest of the intelligence community. Small wonder that outside investigators (and congressional committees) may have an altogether erroneous impression of our current collection capabilities—and thus may tend to blame the analytical process for errors which were only in part its fault.

Some finished reporting of intelligence agencies also contributes to the impression that our knowledge is more current than is actually the case. Order of battle summaries will "accept" the existence or move of a unit months or sometimes even years after it has occurred—but often without mentioning the actual date of the event. Similar time lags in the acquisition and reporting of other important basic data may be well understood by the analysts immediately concerned but not by most readers.

Another, and somewhat more subtle, factor is that the intelligence community from day to day is expected to make judgments about the current situation (including the state of military preparedness or combat readiness) in a variety of countries which habitually conceal or attempt to conceal from us nearly all strategic information. Nonetheless, the intelligence process—looking at the overall situation insofar as it can perceive it and seeing nothing obviously abnormal—concludes that the situation is generally normal, or routine, or that all forces are in their usual locations and in a relatively low state of combat readiness. In fact, such judgments if made on a daily or weekly basis may be based on the most superficial knowledge of what is actually happening at the time and can be quite erroneous. However, the chances are high that such judgments are right—but this will not necessarily be because we know what is happening but rather because about 95 percent of the time things really are pretty normal. Thus, even if no current information were being received, the odds are that the statement would be correct. The impression nonetheless is left, probably in the minds of most readers, that these judgments are the product of considerable evidence. And if we know when things are "normal," then clearly we should also know when something is "abnormal."

Now, there are some types of developments—both military and political—which we have a fair chance of detecting on a current basis in many countries and which, if they occur, can often be regarded as abnormal. In the case of military activities, these will most often be developments, particularly movements, which occur in the forward areas or near the borders where our detection capabilities are usually the best. Certain very obvious political anomalies—such as cancellations of planned trips by the leadership, or extraordinary diplomatic or propaganda developments—also can often be described as abnormal even though their purpose may not be clear. But these obvious developments may be only the smallest fraction of what is really going on, and the activity could well include the initial preparations for future hostile actions—which could be totally concealed.

What we observe from day to day of what goes on in the Communist world—even in those areas in which our collection and basic intelligence is the best—is actually something less than the tip of the iceberg. Disasters—natural and man-made—insurrections, major internal struggles even including the ouster of key officials, mobilization of thousands of men, and a host of other military preparations have been successfully concealed from us and almost certainly will be again. The capabilities of our enemies for concealment are probably fully appreciated or recognized only by a relatively small percentage of those associated with, or dependent upon, the intelligence process. It may take a really surprise hostile action by our adversaries right under our noses for us to realize that fact. But the intelligence system has been burned often enough, and in relatively recent years at that, not to have learned this lesson.

The spectacular aspects of the Cuban missile crisis and the delay in detection of the strategic missiles in Cuba have tended to overshadow the important question of what was going on in the Soviet Union in the spring and summer of 1962 as the USSR was making its preparations for the deployment of the missiles and the accompanying military forces to Cuba. We tend to overlook the fact that our appreciation of what was happening was almost entirely dependent on what we observed on the high seas and in Cuba, and that even today we have only the most limited understanding of the steps which were taken in the Soviet Union itself. A minimum of 20,000 combat troops (and quite probably more) were moved from unknown points in the USSR to Cuba, together with the equipment for entire SAM battalions and MRBM regiments, as well as tanks, short-range missiles and quantities of air, naval and electronic equipment—all without a discernible ripple in the USSR itself. Thousands of troops staged through the Baltic port of Kaliningrad, less than 25 miles from the Polish border, without a rumor of the movement ever reaching the West. Only the fact that this small expeditionary force was then moved by ship permitted us to recognize that anything unusual was under way. Although even by Soviet standards this was perhaps an extraordinary accomplishment in security, it nonetheless should serve as a constant reminder of how little we may know.

A second example of a very well concealed move was the closure of the Berlin sector borders in the early morning hours of 13 August 1961.[1] The warning analyst who often complains that the available indications were overlooked or not appreciated is hard put to find the evidence which was ignored in this case. In fact, the available indications were carefully analyzed and did not *in themselves* support a likelihood that the Soviets and East Germans would move to close the *sector* borders. Such evidence as was available tended rather to indicate that they would seek to cut down the enormous flow of refugees to the West by closing the *zonal* borders (i.e., access to Berlin from East Germany itself), a less drastic but also less effective move which would not have violated the four-power agreements on the status of Berlin. The preparations for the closure of the sector borders were actually carried out almost under the eyes of Western patrols and yet achieved almost total surprise. This success was the more remarkable in that our collection capabilities in East Berlin and through it into East Germany were then considered unusually good—and superior to those in any other area of the Communist world. At all levels, the community suffered from a misplaced confidence that this collection would give us insight into what the Communists were most likely to do—that there would be a leak of some kind as to their plans.

The lessons which the analyst should derive from such experience are at least two:

a. The observed anomalies—those which are apparent—will likely be only a small fraction of the total, and it must be presumed that far more is under way than is discernible to us; and

b. Even in areas or circumstances where collection is very good and we may have much information, our seeming knowledge may be deceptive and the enemy may be capable of far greater concealment than we would normally expect.

Years ago, it was fashionable to have signs hanging in intelligence shops which read, "Do not reason beyond your data." One is happy to note that they have long since disappeared. The intelligence analyst—and above all the warning analyst—must recognize that his data almost certainly will be inadequate. He must be capable of drawing reasonable inferences from that which he does know about that which he does not know.

The Presumption of Surprise

Closely related to the recognition of the inadequacy of our knowledge, and fundamental to the indications method, is the presumption that the enemy usually will attempt to surprise us. If he cannot or does not attempt to conceal completely what he is getting ready to do, he will at least attempt to deceive us on some aspects of his plans and preparations.

It follows that the warning analyst must be inherently skeptical, if not downright suspicious, of what the enemy or potential enemy may be up to from day to day—whether or not there is any great cause at the moment to be particularly worried about his intentions. The indications analyst will examine each piece of unusual information or report for the possibility that it may be a prelude to hostile action or some other surprise move, even though the situation at the moment gives no special cause for alarm. He will not discard information of potential warning significance until he can be sure, or at least reasonably sure, that it is erroneous or that there is some really satisfactory "other explanation" for the anomaly. He will not accept the most reassuring (or least worrisome) explanation for some unusual development which could prove to be ominous. He will endeavor consciously to go through this analytical process and to maintain his alertness for unexplained anomalies—and he will hold on to those fragments of information which are potentially of indications significance.

This approach is thus somewhat different—in some cases markedly different—from that of the current or basic analyst, or even from that of the estimator, though all are using the same basic information. It is the function of the warning or indications analyst to be alert for the *possibility*—however remote it may seem now—that the enemy has begun preparations for hostile action. At least in theory, the indications analyst will be running ahead of the rest of the analytical community in his perception of this possibility. It is his function, if the system operates as it should, to be continually raising questions concerning the enemy's possible motives, to be reminding others of pieces of information which they may have overlooked, to be urging more collection on items of possible warning significance. He is the devil's advocate, the thorn in the side of the rest of the community. This method is sometimes also described as assuming "the worst case."

There has been much misunderstanding about this, and many have derided indications analysts for their presumed proclivity for crying "wolf! wolf!" Warning intelligence is held in ill repute by some—often those with the least contact with it—because they assume that indications analysts are continually putting forth the most alarmist interpretations, in part to justify their existence or because that is what they feel they must do in order to earn their pay. Thus the indications analyst *proposes* but fortunately wiser heads prevail and the current intelligence or estimative process *disposes* and puts those scaremongers in their place.

It is necessary emphatically to dispel the idea that the perceptive and experienced warning analyst is continually rushing to his superiors with the most alarmist interpretations or is an irrational and undependable character. No responsible indications analyst takes the "worst possible" view of every low-grade-rumor or is continually searching for the worst possible explanation of every anomaly. Rarely if ever will he regard a single report or indication as a cause for alerting the community. To maintain an open and skeptical mind and to be diligent and imaginative in the collection and analysis of evidence against the possibility of the unexpected does not require that one go off half-cocked.

Indeed, experience has shown that qualified warning analysts, well-versed in their facts, often have been able to play the opposite role—they have sometimes been able to dampen down potential flaps engendered by a few alarmist rumors or unconfirmed reports, particularly when the international situation is tense or a general crisis atmosphere prevails. One product of the diligent collection of facts is that the analyst is able to make a reasoned and hopefully more objective analysis than he would otherwise be able to do.

The Scope of Relevant Information

As a product of the diligent collection of facts, and possible facts, the indications analyst hopefully should be able to assemble and analyze *all* the available information which may bear on the problem and not just that received most recently or that which is most apparent or readily acceptable.

Warning intelligence must deal not only with that which is obvious but with that which is obscure. It must consider all the information which may be relevant to the problem at hand. If one accepts the premise of surprise, it will follow that what the enemy is preparing to do will not necessarily be that which is most obvious or seemingly plausible.

We have earlier (in Chapter 3) discussed the importance for warning of the exhaustive research effort and the requirement that the analyst have available the basic raw data. Again, the importance of this can hardly be overemphasized. We have also (in Chapter 8) discussed the preparation of chronologies as a method whereby the analyst can assemble in readily usable form a variety of reports and fragments of information which may relate to the problem even though their relevance cannot yet be positively established.

These tools of analysis are invaluable; they cannot be dispensed with. They permit the analyst—assuming that the system has functioned as it should and that he really has at hand all or nearly all the potentially relevant data—to begin to fit his pieces of information together into some meaningful pattern, to check out various hypotheses against the available data. But these tools will not, in and of themselves, solve the problem of relevancy. They will assist the analyst to think and to prepare an argument which may be convincing to others, but rarely will such techniques provide "proof" that a given piece of information is pertinent or is something which should be considered in reaching an assessment of the enemy's intent. In the end, the question of what is relevant and what is not is a problem for human judgment.

It is a characteristic of impending crises or of periods of great national decisions and extraordinary preparations that there are likely to be a large number of unusual developments and often a still larger number of unconfirmed, unexplained and otherwise puzzling reports or rumors. Some of these will be true but unimportant. Some will be false but—had they been true—potentially important. Some will be true, or partly true, and very important. But a lot of these reports simply cannot be determined with any certainty to be either important or unimportant. It cannot be demonstrated with any certitude whether they are or are not relevant. And, interestingly enough, this may never be established. Time will neither prove nor disprove the relevance of some information. Which is one reason that postmortems, even with all the benefit of 20-20 hindsight, often do not reach agreements either.

The problem of what facts, what known occurrences, what reported developments, which rumors, and how much of what is happening in general are actually pertinent to what the enemy may be getting ready to do is one of the most difficult problems for warning intelligence. It is probably safe to say that no two people are likely to reach complete agreement on what is relevant in any really complicated situation. One of the most frequent criticisms levied against the indications method is that it tends to see all unusual or unexplained developments not only as potentially ominous but also as related to one another. This same criticism is often made against chronologies on the grounds that they tend to pull together a lot of developments or reports simply because they occurred at the same time when there is actually no demonstrable causal relationship between them. Obviously, such criticisms can have much validity if in fact the indications analyst has been tossing in everything but the kitchen sink in an effort to demonstrate ominous connections between a variety of reports and developments of uncertain validity and/or significance.

Nonetheless, in justice to the indications method, it must also be said that in retrospect—when the crisis is over—more rather than less information is nearly always judged to have been relevant than was appreciated by most people at the time. The imaginative rather than restricted or literal approach has almost invariably proved to be the correct one. The ability to perceive connections, or at least possible connections, between events and reports which on the surface may not seem to be directly related is a very important ingredient in the warning process, and one which has probably been given too little attention. The formal process of intelligence and the emphasis on coordination and "agreed positions" have tended increasingly to suppress the independent and often more imaginative analysis. "Think pieces" and speculative analyses need to be encouraged rather than discouraged. Information which may be relevant to the problem, whether or not it can be "proved" to be, must be considered in coming to warning assessments. It was the inability to see the relevance and interconnections of events in the Soviet Union, Germany and Cuba which contributed in large part to our slowness to perceive that the USSR was preparing for a great strategic adventure in the summer of 1962.

When we are attempting to assess what may and may not be relevant in a given situation, it may be well to be guided by an important and usually valid precept. As a general rule, the greater the military venture on which any nation is preparing to embark, and the greater the risks, the more important the outcome, the more crucial the decisions—the wider the ramifications are likely to be. And the wider the ramifications, the more likely it is that we will see anomalies in widely varying fields of activity as the nation prepares for the great impending showdown. In these circumstances, many seemingly unrelated things really do have relevance to the situation—if not directly, then indirectly. They are part of an atmosphere; they contribute to the sense of unease that things just are not right or normal, and that something big is brewing. To discard a series of such fragments as irrelevant or unrelated is to lose that atmosphere. The intelligence system will usually come out ahead if it is prepared in these circumstances to consider as relevant a broader field of information than it might in other circumstances. It will be better prepared to accept the great dramatic impending event if it has already perceived a series of anomalies as possibly relevant to such a development. In the words of Louis Pasteur, "Chance favors the prepared mind."

Objectivity and Realism

No factor is more important for warning than objectivity in the analysis of the data and a realistic appreciation of the situation as it actually is. The ability to be objective, to set aside one's own preconceptions of what the enemy ought to do or how he should behave, to look at all the evidence as realistically as possible—these are crucial to indications analysis and the ultimate issuance of the warning judgment at every stage of the process. The greater experience any individual has in the warning field the more likely he is to believe that objectivity, and the accompanying ability to look at the situation as the other fellow sees it, is the single most crucial factor in warning. There have been too many warning failures which seemed attributable above all to the failure of people—both individually and collectively—to examine their evidence realistically and to draw conclusions from it rather than from their subjective feelings about the situation.

One of the foremost authorities on warning in the US Government, when told by this writer that she was planning this book, said, "So what have you got to write about? It is the same thing every time. People just will not believe their indications."

The rejection of evidence incompatible with one's own hypotheses or preconceptions, the refusal to accept or to believe that which is unpleasant or disturbing or which might upset one's superiors—these are far more common failings than most people suspect. One of the most frequent—and maddening—obstacles which the warning analyst is likely to encounter is the individual who says, "Yes, I know you have all these indications, but I just do not believe he'll do that." Korea, 1950; Suez, 1956; Hungary, 1956; Czechoslovakia, 1968. In none of these cases did we lack indications—in some of them, there was a wealth of positive indications—it was that too many people could not or would not believe them.

Now, the reaction of most people to the mere suggestion that they may not be thinking objectively is one of high indignation—it is an insult both to their intelligence and their character to imply such a thing. Of course, they are thinking objectively; it's you or the other fellow who is not, or it is you who are leaping to conclusions on the basis of wholly inadequate evidence while they are maintaining an open mind. In these circumstances, the atmosphere often becomes highly charged emotionally. Positions tend to harden, and with each side taking a more adamant position, reconciliation of views becomes less likely, and objectivity more difficult to obtain.

One of the first things which we all must recognize if we are to make progress in the warning business is that *nobody* achieves total objectivity. We are all influenced in some degree by our preconceptions, our beliefs, our education, early training, and a variety of other factors in our experience. Some people—and for reasons yet to be fully understood—are capable of more objectivity than other people. Or, perhaps more accurately, they are capable of more objectivity on some subjects which other people find it difficult to be objective about. But no one is perfect.

We must also recognize that the ability to think objectively is not necessarily correlated with high intelligence, and that objectivity in one field may not necessarily carry over into other fields. The history of both religion and of politics provides ample illustrations of men of great intellectual achievements who were nonetheless totally biased and dogmatic in their outlook. Cases can be cited also of brilliant scientists—presumably capable of very objective analysis in their fields of specialization—who have seemed almost unbelievably naive or unrealistic when they have attempted to engage in other pursuits. Whole communities and even nations (Salem, Massachusetts, and Nazi Germany) have been victims of mass hysteria or guilty of such irrational conduct that it is difficult for the outsider to comprehend it at all. None of us is immune. We are all emotional as well as rational.

It is important, both for the analyst and the intelligence community as a whole, to recognize and face this problem. It should not be a forbidden subject or no progress will be made. The rules of the game generally have precluded challenging a colleague publicly, and certainly not one's superiors, as to *why* they will not accept certain evidence or *why* they think as they do. Yet this may be the most crucial factor in their assessment. It would help greatly in such cases if people could be induced to explain their thought processes as best they can, or what is really behind the way they feel. The mere process of discussing the subject, dispassionately one would hope, may help them to see that they are not really assessing the evidence on its merits, but evaluating it in the light of their preconceptions. Still better, other listeners less involved in the problem may perceive where the difficulty lies.

For years, experienced indications analysts have maintained, and not facetiously, that warning was a problem for psychologists. What happens to some people that they cannot examine the evidence with greater objectivity, that they will consistently reject indications which seem self-evident to another analyst? Why will one analyst view certain information as tremendously important to the problem and another tend to downplay it as of not much importance or discard it altogether? How can there be such a wide divergence of views on the same information—not little differences, but 180 degree differences?

These feelings that psychological, rather than strictly intellectual, factors are involved in warning have recently had support from the academic community. A number of studies by social scientists have inquired into the formation and nature of individual and group beliefs and their relationship to national decision-making. Some conclusions which have emerged from these studies are that:

- Individuals do not perceive and evaluate all new information objectively; they may instead fit it into a previously held theory or conceptual pattern, or they may reject the information entirely as "not relevant."
- If the individual has already ruled out the possibility of an event occurring, or considers it highly unlikely, he will tend also to ignore or reject the incoming data which may contradict that conclusion.
- A very great deal of unambiguous evidence is required to overcome such prejudgments or to get the analyst to reverse his position even to the point of admitting that the event is possible, let alone probable.
- If the individual is unable to assimilate this contradictory information into his existing frame of reference or cannot be brought to modify his opinion, the extreme result may be a closed-mind concept of the situation with a high emotional content.[2]

All of the foregoing reactions have been observed in warning situations. It is unquestionably true that this is precisely what has happened in more than one crisis, to the detriment of the issuance of warning.

To recognize that a problem exists at all is the first step toward attempting to solve it. If we can accept that objectivity is a real and very serious problem in warning situations, and that lack of objectivity can affect the Chiefs as well as the Indians, we may begin to cope with it. If this handbook does nothing more than to teach this precept, it will have served its purpose. In fact, much of the rest of this treatise is designed to help the analyst, his supervisor and the policy official as well to analyze, evaluate and come to judgments that are more objective and hence more accurate.

The Need to Reach Conclusions NOW

The problems of indications analysis, which at best are complex, are immeasurably compounded by the requirement to reach conclusions or judgments long before all the evidence is available or can be adequately checked, evaluated or analyzed. The warning analyst usually does not have the luxury of time, of further collection and analysis, of deferring his judgment "until all the evidence is in." In many ways, he must act in defiance of all that he has ever been taught about careful research—to be thorough, to wait until he has looked at everything available before making up his mind, to check and recheck, to take his time, to come to the "definitive" rather than the hasty judgment. The more extensive his academic training, the more he may be dedicated to these principles or habits, and the more difficult it may be for him to revise his methods when confronted with the real life of current or indications intelligence. Some analysts are never able to make this adjustment. They may make good analysts on some long-term aspects of basic intelligence, but they should not be assigned to the indications field.

The requirements of warning analysis—for both the individual and the system as a whole—in fact may seem to present almost irreconcilable contradictions. The analyst and the community are confronted with the requirement to come to a positive judgment on the elusive and complex problem of the actual intentions or impending course of action of an enemy nation, which may be attempting by all means at its command to conceal that intention or course of action. And, lest it be too late tomorrow to do so, the analyst must come to some judgment today even though the evidence is clearly inadequate, highly conflicting, and the relevance of some of it (even if its reliability could be established) unknown or subject to differing interpretation. If it is not too late tomorrow to issue the warning, there will likely be more information at hand—but this in turn may not necessarily clarify the situation or permit a firmer judgment. It may only serve to introduce new factors or considerations and further compound the doubt and confusion. On the other hand, it could lead to a markedly different conclusion, either positive or negative, from the day before. The situation probably is indeed highly fluid.

The reluctance of people, both as individuals and as members of a group, to come to positive judgments in such situations can well be understood. It is not just a concern that a judgment reached today may be refuted tomorrow, nor can it be ascribed just to timidity and fear of being wrong—a matter addressed at greater length in the chapter, Reaching and Reporting the Warning Judgment. The analyst, despite every effort to be objective, may be genuinely uncertain and confused as to what the available evidence means. He really may not know what to think in the face of information so conflicting, tenuous, and elusive. Further, his opinions may be subject to marked fluctuations almost from hour to hour as more information becomes available.

The tenuousness of it all, the uncertainties, the doubts, the contradictions, are characteristics of every true warning problem. Only in retrospect can the relevance, meaning and reliability of some information ever be established, and some of it (often a surprising amount) is never established. In retrospect, however, much of the uncertainty is either forgotten, or it seems to disappear.

Even those who actively worked the problem tend, after the event, to think that they saw things more clearly than they really did at the time. And outsiders—including those assigned to do critiques of what went wrong or to investigate the intelligence "failure"—can never truly see how complex and difficult the problem was at the time.

It is simply impossible (and the word is deliberately selected here) to recover in retrospect all the thought processes of those who participated in the analysis and assessment in a true warning crisis. The situation can never truly be reconstructed or relived. Post-mortem reexaminations of the information which was at hand, particularly if done in great detail from the original source material and administrative logs, can help, but they cannot really recapture the atmosphere, Only if everyone associated with the problem kept a completely honest diary, almost hour by hour, could the intellectual and emotional experience be fully recovered. This has been done in part, however, for some crises. The student of warning is referred in particular to some of the published history of the Cuban missile crisis which at least gives some sense of the tremendous uncertainties which plagued both the intelligence and policy levels of the government.

Some appreciation of how it really is in a crisis—however inadequately it has here been described—may help to give the analyst new to the business a little more confidence in himself. The fact is that if you are doing a good job you will have some doubts and uncertainties. No one who has tried to look objectively at all the available evidence can come to judgments with full certainty—there s bound to be some doubt, if only about those things you do not know.

Excessive certainty (or seeming certainty) in these circumstances is likely the mark of the closed mind—of those who will not examine the evidence objectively lest it upset their already preconceived hypotheses.

If the analyst has doubts and uncertainties, that's normal. If he is nonetheless willing to come to a judgment as best he can, to put his money on one probability or another, he probably has some aptitude for warning analysis. If he cannot do this—if he must forever "reserve his position"—he would better be assigned to some other field of endeavor.

Notes

1. Contrary to widespread popular opinion, the Berlin wall was not constructed that night; it was not even begun until about 10 days later. The sector borders were initially closed by road blocks, barbed wire and troops.

2. A very good, brief summary of these studies was contained in an article by Loren R. Larson entitled "Objectivity in Intelligence," published in the DIAAP Journal of Intelligence January 1970.

11

Some Specifics of the Analytical Method

Inference, Induction and Deduction

As SHOULD BE EVIDENT from the discussion this far, the analysis of indications and the reaching of the warning judgment almost always will be a process of inference. Also, it will in large part be the result of inductive rather than deductive reasoning. Or, to simplify what is a very complex problem, the warning judgment will be derived from a series of "facts" or more accurately what people think are the facts, the inferences or judgments which may be drawn from these conceptions of the facts, to a final conclusion or series of conclusions which will be expressed as probabilities rather than absolutes. The process is highly subjective; neither the inferences nor the final conclusion follow necessarily from the facts and are therefore not provable, and, as a result, individuals will vary widely in their willingness to reach either specific inferences or general conclusions.

This is not a chapter in the nature of logic, and for our uncomplicated purposes we may use the relatively simple dictionary (Webster's) definitions:

Inference: the act of passing from one or more propositions, statements or judgments considered as true, to another the truth of which is believed to follow from that of the former;

Induction: reasoning from a part to a whole, from particulars to generals, or from the individual to the universal;

Deduction: reasoning from the general to the particular, or from the universal to the individual, from given premises to their necessary conclusion.

From the nature of the evidence which will be available to the warning analyst, he will normally have no choice in the warning situation. He must proceed from fragments of information and from particulars to general conclusions, and he will not have given premises which will lead logically or necessarily to certain conclusions. He must normally come to his judgments—both as to the facts in the case and their meaning—on the basis of very incomplete information, on a mere sampling of the "facts," and often without knowing whether he even has a sampling of some of the potentially most important data. Unlike the scientist who can select his cases and who will usually be able to judge whether he has a "statistically significant" sampling, the indications analyst must work with the samples or crumbs which fall his way and which may or may not be representative of a broader trend. Nonetheless, he must do the best which he can with the data at hand since, as we have noted, time may not be on his side.

The problem of how much evidence is necessary in order to come to certain inferences or to some type of generalized conclusions is peculiarly applicable to certain types of military information—above all, to order of battle and mobilization and these problems are discussed in greater detail in the next major section (Specific Problems of Military Analysis). For purposes of illustration, however, we will note here that there is a fundamental difference in analytical approach between the conventional order of battle method and the indications method.

The normal order of battle approach to assessing the strength and locations of enemy forces is to examine each unit individually. Ideally, the order of battle analyst seeks the total sample before he wishes to make a judgment on total strength. He usually requires positive evidence for each individual unit before he is willing to make a judgment (to accept) that the unit has been reequipped with more modern tanks or higher performance aircraft, or that it has been mobilized with additional men and equipment to bring it to wartime strength. The conventional military analyst normally will be extremely reluctant to accept that any new military units exist or are in process of formation until he has positively located them and identified them until they meet some type of order of battle criteria. This is likely to be the case even if there is a great deal of other information to indicate that a major mobilization of manpower is in fact in process. The same general precepts apply to movements of units. It is not enough that there are numerous reports of movements of ground units and aircraft; what the order of battle analyst wants is evidence that this or that *specific unit* has moved. His method is thus essentially restrictive rather than expansive. He will not normally conclude that more units have moved than he can identify, or that several units have been mobilized because he has evidence that one or two have.

The indications analyst, on the other hand, must consider that what he sees or can prove may be only a fragment of the whole and that he may have to come to more general conclusions about what the enemy is doing from his sampling of the evidence. If available information—unspecific though it may be—suggests that large-scale callups of men to the armed forces are in progress and if some units which he can observe are known to be mobilizing, the indications analyst will consider it a good chance, if not certain, that other units also are being mobilized. If one unit in the crisis area has issued the basic load of ammunition, it will be likely that others have. If some units are known to be deploying toward the front or crisis area but the status of others is unknown, the indications analyst will be inclined to give greater weight to the possibility that additional units are deploying than will the order of battle analyst. At least he will consider that what is provable is probably the minimum that is occurring, rather than the maximum.

The warning analyst does not take this approach because he wants to draw broad judgments from inadequate data or because he is inherently rash and likes to jump to conclusions. He takes this attitude because this is the nature of warning problems. In times of crisis, the more leisurely and conventional analysis may be a luxury which we cannot afford. Much as we might wish to have all the data before we draw our conclusions, we may have to make general judgments based on only a sampling of the potentially available data, or whatever we can get. How great a sampling is needed before a general conclusion can be drawn will vary with the circumstances and is likely in any case to be a hotly debated subject between warning analysts and more conventional military analysts. The point here is that the principle of inductive (or more expansive or generalized) analysis must be recognized as valid in these circumstances, even though judgments will necessarily be tentative and subject to some margin of error. Historically, as we have noted in Chapter 5, intelligence has usually underestimated the scale of mobilization and troop deployment in a crisis situation.

The problems of reaching general judgments from limited data are of course applicable to other types of information which are received in a potentially critical situation. Even assessments of the meaning and significance of political information or propaganda trends must be based on a sampling of what is actually occurring and are essentially inferential or inductive in nature. Because so much political planning and preparation may be concealed, however, and because other political evidence may be so ambiguous or uncertain, political analysis in a crisis situation may be even more complex and subjective than the ascertainment of the military facts.

The Acceptance of New Data

It is a phenomenon of intelligence, as of many other fields of investigation and analysis, that the appearance of new types of information or data of a kind not normally received poses difficult problems—and that the reaction is likely to be extremely conservative. This conservatism, or slowness to accept or even to deal with new information, is a product of several factors. One is simply a basic tenet of good research—that judgment should be withheld in such cases until sufficient unambiguous data are available that we can be sure of the meaning and significance of the information. A related factor often is that there is no accepted methodology for dealing with new types of information; since it is new, the analyst is not sure how to tackle it analytically, and he wants more time to think about it and to confer with other analysts. Still another and often very important factor may be that the intelligence organization is not prepared to handle this type of information or analysis; there is no one assigned to this type of problem, so that it tends to be set aside or its existence may not even be recognized. And finally, there is an inherent great reluctance on the part of many individuals and probably most bureaucratic organizations to stick their necks out on problems which are new, controversial, and above all which could be bad news for higher officials and the policy maker.

The effect of these factors and possibly others, individually and collectively, can be to retard the analysis and acceptance of data in the intelligence system by weeks, months and sometimes even years. Few people outside the analytical level have probably ever recognized this, but examples can be cited of delays in the analysis and reporting of data which seem almost incomprehensible in retrospect. Even individuals normally extremely perceptive and imaginative seem to have lapses in which they are unable to perceive that something new is occurring or that things have changed. They may be reluctant to accept evidence which in other circumstances—particularly when there is adequate precedent—they would accept without question.

Lest the writer be accused of exaggerating all this, a few examples may be in order.

A very intelligent and sophisticated supervisor of political intelligence on the Soviet Union resisted for weeks (and finally took a footnote in opposition to) a judgment that the USSR was preparing in the mid-1950's to export arms to the Middle East. His reason—although the evidence was almost overwhelming and no one would question it today—was that the USSR had never exported armaments outside the Communist world, and therefore never would.

Numerous examples could be cited of delays in the acceptance of information on troop movements, other order of battle changes and mobilization, which were attributable to the fact that the evidence was new rather than insufficient by criteria later adopted. Since the order of battle problem is dealt with in greater detail in Part IV, two instances from the Vietnam war may suffice here:

It required ten months for order of battle to accept the presence in South Vietnam of the first two North Vietnamese regiments which arrived there in the winter of 1964–1965 (one other which arrived slightly later in February was, however, accepted in July). With the acceptance at the same time of several other regiments which had arrived in 1965, and with it the acceptance of the fact that North Vietnam was deploying integral regiments to South Vietnam, the time lag between arrival and acceptance thereafter rarely exceeded two months and was often less.

North Vietnam began an expansion of its armed forces in 1964 and conducted a major expansion and mobilization in 1965. Much of the evidence for this mobilization was derived from public statements and other indirect evidence of large-scale callups to the armed forces, rather than from conventional order of battle information. (Confirmation by order of battle ran months and in some cases years behind what could be deduced from what Hanoi was saying about its mobilization effort.) Just why there was so much reluctance to accept Hanoi's transparent statements that thousands (sometimes tens of thousands) of men were required to fight at "the front" (i.e., in South Vietnam) was then and is now very difficult to understand. Overreliance on OB methodology and mistaken notions that anything said by the enemy should be dismissed as "propaganda" no doubt contributed. In time, however, the value of Hanoi's press and radio comments came to be recognized. A few years later, public statements that callups to the armed forces were being increased were routinely accepted as evidence that this was in fact under way, whereas much larger volumes of similar evidence were not considered acceptable in 1965–66. (See more detailed discussion of the North Vietnamese mobilization problem in Chapter 15.)

These examples are cited for purposes of illustration only. Many other instances, some probably more serious, could undoubtedly be found from the annals of warfare. The fact is that the reluctance to cope with or accept new types of data is a very serious problem for warning intelligence, and one to which supervisors in particular should be constantly alert. A review of draft indications items, when compared with what was finally accepted for publication, will consistently show that indications analysis tends to run ahead (sometimes weeks ahead) of what is usually acceptable to the community as a whole, particularly on new intelligence. One is happy to note, however, that this gap appears to be narrowing—which may reflect somewhat greater willingness to use indications methodology and new types of data than was once the case.

Together with the need for objectivity, the ability to perceive that change has occurred and that new information must be considered even though it may not yet be fully "provable" is a fundamental requirement for successful warning analysis. In warning, delay can be fatal. Both the analyst and the system as a whole must be receptive to change and the acceptance of that which is new.

Interestingly enough, experience suggests that experts in their fields are not necessarily the most likely to recognize and accept changes or new types of data. The greatest resistance to the idea that Communist China would intervene in Korea in 1950 came from "old China hands," and by and large it was Soviet experts who were the most unwilling to believe that the Soviet Union would put strategic missiles in Cuba. There seems to be something to the concept that experts tend to become enamored of their traditional views and, despite their expertise, are less willing to change their positions than those who are less involved with the matter. One student of this subject has concluded that scholars who study the Soviet Union using a theoretical model are less likely to see evidence for change than if they had no model. He concludes:

> It is no accident . . . [that] American journalists and casual observers are more likely to see change in the Soviet Union . . . than are the so-called "Soviet experts". . . . We must entertain seriously the . . . possibility that the "expert" tends to defend his own theoretical position by virtue of the selective interpretation of ambiguous data.[1]

Understanding How the Enemy Thinks

This heading may be a bit misleading. No one can aspire to total understanding of how someone else, and particularly the leadership of another and hostile nation, actually thinks. A sophisticated student of the Soviet Union (the remark is attributed to Chip Bohlen) once observed that his favorite last words were: "Liquor doesn't affect me," and "I understand the Russians." To which the warning analyst should add a fervent, "I understand the Chinese."

The path to understanding the objectives, rationale, and decision-making processes of foreign powers clearly is fraught with peril. Nonetheless, it is important to *try*. The analyst, the intelligence system and the policy maker or military planner may have to make a conscientious and imaginative effort to see the problem or situation from the other side's point of view. Fantastic errors in judgment, and the most calamitous misassessments of what the enemy was up to have been attributable to such a lack of perception or understanding. An examination of such errors in perception—both by individuals and groups—which have been made over a period of years in warning situations indicates that this is a problem to which we must be particularly alert.

The ability to perceive, or to attempt to perceive, what others are thinking may be the mark of an expert in more than one field. A prominent bridge columnist has observed that an important attribute of the expert bridge player is his "faculty of being able to capitalize on his knowledge of what the opponents are thinking. That is, he views some specific-situation not only through his own eyes and mind, but also through the eyes and minds of his opponents." This ability does not seem to come easily to some people—even when the opponent is making no particular effort to conceal how he feels about the matter, and indeed may be making it quite obvious. Why did so many analysts—after months of evidence that the Soviet Union was determined to maintain its political hold on Czechoslovakia and in the face of a massive military buildup—nonetheless bring themselves to believe that the Soviet Union would not invade? One can only conclude that they had not really tried to understand how the USSR felt about it, how important the control of Eastern Europe was to Soviet leaders, and that these analysts had somehow deceived themselves into believing that a detente with the US was more important to the USSR than its hegemony in Europe and the preservation of the Warsaw Pact.

The root causes of such lack of perception—whether it be an inability or unwillingness to look at it from the other fellow's standpoint—are probably complex. Individual, group and national attitudes or images are involved as well as relatively more simple questions such as how much education the analyst has on the subject, how many facts he has examined, and how much imagination he has. The problem of objectivity and realism obviously is closely related; misconceptions, based on subjective judgments of how the other nation *ought* to behave rather than objective assessments based on how it *is* behaving, have much to do with this problem.

While one hesitates to suggest that there may be certain national characteristics of Americans which tend to inhibit our understanding of what our adversaries may be planning, one is struck by two prevailing attitudes:

Perhaps because of our historic isolation, prosperity and democracy, Americans traditionally have been optimistic, and often unduly so, about world affairs. A reluctance to believe that World War II would come or that we could become involved lay behind much of the isolationist sentiment of the thirties. We couldn't believe that Japan would be so treacherous as to attack Pearl Harbor, that the Chinese would intervene in Korea, or that our close friends the British would move against Suez in 1956. And so forth. Healthy and admirable as such attitudes may be in private life, we need to guard against such false optimism in professional intelligence analysis.

Many Americans also—and this has sometimes been particularly true of military men—are so convinced of the superiority of American military forces and technology that they cannot bring themselves to believe that more backward and ill-prepared nations would dare to oppose US military power. Perhaps more than any other factor, this lay behind the unwillingness to believe that Peking would throw its poorly equipped and ill-educated troops against the armed forces of the US in Korea. Much of the same attitude prevailed in the early days of the Vietnam war, although probably no longer.

Irrational and misguided as it may seem to us, people often will fight for national causes and objectives in the face of seemingly superior forces, including ours. And great nations will overwhelm little nations. We would do well to recall our past mistakes and to remember that understanding how the enemy thinks, what is important to him and what he will fight for can sometimes be more crucial to warning than the most detailed understanding of the locations and capabilities of his military units.

This perception of the national objectives and priorities of other countries—and of the strategic importance to them of a particular issue—is so fundamental to political warning that this subject is addressed at greater length in the section on political warning (particularly Chapters 19 and 20).

The Consideration of Various Hypotheses

The consideration of alternative or multiple hypotheses to explain sets of data is a fundamental of the scientific method which, curiously enough, often is given scant attention in intelligence problems. Various alternative explanations or possibilities may be offered for particular facts or bits of information—e.g. this photography of new construction activity could be a missile site in its early stages but it may be an industrial facility. Often, however, no particular effort will be made to itemize all the available pieces of information on any complex current situation with a view to considering their relevance to various alternative possible courses of action of the adversary. Even special national intelligence estimates (SNIEs), prepared sometimes to address alternative possibilities, usually do not attempt to deal with all the facts or possible facts before coming to judgments. They are much more apt to be generalized discussions of alternatives, rather than detailed analyses of the relevant information. There have been certain exceptions to this—on particularly critical military subjects, for example—but as a general rule estimates do not involve a critical examination of a mass of detailed information and a consideration of various hypotheses to explain it. Still less are other forums of the intelligence community likely to provide an opportunity for detailed consideration of various alternative hypotheses unless a special effort is made to do so and the forum is opened to all who may have something to contribute.

Ideally, the warning system should operate to provide a forum for this type of analytical effort—but often it does not. The analysts who have the most detailed knowledge should but often do not get together with analysts of other agencies on warning problems. More often, each agency considers the evidence and comes to its "position" and the various "positions" are what are discussed at the inter-agency forum rather than the detailed evidence itself. The result is that various alternative hypotheses may not be given adequate consideration, or even sometimes considered at all, and no systematic effort is made to ensure that some group really goes through *all* the evidence and considers the various alternatives explanations in exhaustive detail.

One reason for this—which we have noted before—is that in crises, or budding crisis situations, there is likely to be an overwhelming quantity of information, the mere scanning and preliminary processing of which is consuming most of the analysts' time. There are simply insufficient resources to cope with all the information in any manner, let alone go through a time-consuming process of evaluating each item of information against several alternative hypotheses.

This is a serious matter. Most of all, in such situations, people need time to think—to really take the time and effort to look to the available information and to consider what it does mean, or could mean. As we have observed in preceding chapters, facts are lost in crisis situations, and sometimes very important facts whose adequate evaluation and consideration would have made a great difference indeed to the final judgment.

The most useful thing which an administrator or committee chairman can probably do in such circumstances is to devise some method to ensure that all the relevant information, or possibly relevant information, is being brought together and that it is really being looked at and considered against various hypotheses or possible courses of action. Before the validity of various hypotheses can be considered, we must ensure the examination of the facts.

Assuming that this can be accomplished, an objective consideration of the meaning and importance of individual pieces of information in relation to various alternative hypotheses can be a real eye-opener. Almost any method which will require the analyst or analysts collectively to examine the data and evaluate each piece will help. One does not need for this purpose some mathematical theorem or other statistical method to ensure "reliability" or "objectivity"—the purpose is to look at the information and to say yes or no or maybe to each piece (to each indication) as to whether it is or is not a likely preparation for (a) hostilities, (b) peace, (c) various stages between full-scale war and total peace, of which of course there may be many, or (d) not significant for any of these hypotheses.

The idea that the analyst should be required to look at each piece of information and to come to a judgment on *it* and *its* validity or relevance to various hypotheses is the essence of various systems or theories which have been devised for the purpose of helping to improve objectivity in analysis of information. The best known of these systems, which enjoys considerable popularity today in the information handling field, is "Bayes Theorem." (We will be considering this and other related techniques in a later chapter entitled "Assessing Probabilities.") In Bayes Theorem, the analysis or judgment concerning each piece of information is but the first step—the analyst then applies a likelihood ratio to each item and the results are mathematically evaluated to come to an overall probability or likelihood ratio for a given hypothesis. The theory behind the theorem is that the analyst is more capable making an objective judgment concerning a given piece of information than he is in coming to an overall objective conclusion on the basis of many pieces of information, and that the method will ensure greater objectivity of the conclusion.

Bayes Theorem may or may not have merit in warning situations—the writer is in no position to say but understands that it has yet to be demonstrated that this is so and that the system, as applied to warning, is still experimental. The proponents of the system, however, have overlooked their trump card—which is that the mere setting up of hypotheses and requiring the analyst or analysts to evaluate information against the hypotheses is in itself an enormously important step. If this could be done in all crisis situations—without regard to whether any mathematical formula was then applied—the intelligence system would come out ahead. For the merit of this system—or any similar system—is that it induces the analyst to think, and to evaluate, and not to ignore information which he finds distressing or contradictory to his going hypothesis.

The outstanding successes of warning analysts, and of watch committees, have usually been those which involved the meticulous and objective evaluation of each piece of information and its relevance to war or peace. Any system which requires people individually and collectively to apply some such quasi-scientific method to their data and their analysis is almost certain to be of positive benefit to warning. A distinguished former chairman of the US Watch Committee repeatedly observed that what the report included and how it was organized was infinitely less important than that the committee painstakingly examine each potential indication and evaluate it.

The advantage of such methods is not just to ensure that facts are not ignored, valuable as that is. An even more important result should be that certain hypotheses are not ignored or swept under the rug—particularly those which are frightening, unpopular, or counter to the prevailing mood or "climate of opinion." It is essential for warning that the intelligence system be able objectively to consider the hypothesis which is contrary to the majority opinion or which runs counter to hitherto accepted precepts or going estimates. In addition to an extraordinary analytical effort, this may require an exceptional degree of objectivity and willingness to consider the unpopular thesis, and the one which might require us to take some positive and difficult or dangerous action.

For warning, we rarely need to be concerned about the "idea whose time has come." It is the fate of the idea whose time has not yet come—the hypothesis which is in its infancy and has yet to gain adherents—that should most concern us. There is inertia to new ideas; long-held opinions are extremely slow to change except in the face of some extraordinary development or unambiguous evidence. This is true even when the issues are not important. When the matter is exceptionally important and the new judgment is unpopular or contrary to going national policy, inertia to new ideas or hypotheses may change to outright hostility. The warning system must ensure that this does not happen and that new hypotheses and ideas are given "equal time" on the basis of their merits, no matter how unpopular or contrary to prevailing opinion they may be.

Note

1. Raymond A. Bauer, "Problems of Perception and the Relations Between the United States and the Soviet Union," *Journal of Conflict Resolution* Vol. V, No. 3, September 1961, p. 227.

12

What Makes a Good Warning Analyst?

It is with some hesitance that one approaches this question, since so far as the writer knows there have never been any studies on this subject, nor is there any literature addressed to this specific topic. Psychologists and other behavioral scientists, however, have begun increasingly to inquire into how people think and the extent to which their basic intellectual and psychological attitudes are fixed early in life or may be modified. There would probably be much room for further research on the applicability of many of these studies to the selection and training of intelligence analysts in general and to warning analysts in particular.

Much of what follows in this chapter will not be based on such studies, however, although some reference will be made to them. Most of this discussion derives from the observation of the attitudes and performance of many people in many crises or potential crises. Thus, although these opinions could be called subjective and no doubt will be open to some dispute, they are not intuitive so much as the product of experience.

Anyone who has been an analyst in the field of intelligence, or any other comparable intellectual pursuit, will usually be able to identify other analysts in whom he has confidence, the ones who can be depended on for hard work and careful judgments, and he can tell the "good guys" from the "bad guys." (Hopefully, the supervisory level will also be able to make such assessments!) It will not be unlikely that many of the following intellectual and personal attributes would be considered important for any field of research; some, on the other hand, would seem particularly applicable for warning.

Basic Intellectual Attributes

Among a variety of intellectual attributes which the warning analyst (and his supervisor) should have, it may be possible to single out a few as of particular importance:

a. An insatiable intellectual curiosity. Perhaps more than any field of intelligence research, indications analysis involves the piecing together of fragments of information of uncertain reliability and questionable meaning. Unless the analyst is fascinated by the problems of the unknown, unless he regards each new fragment as a challenge to his ingenuity, and unless he cannot rest so long as there are unresolved problems or missing links—then he is likely to find indications research tedious rather than fascinating. Warning analysis is for seekers after the truth who are unsatisfied with the ready or easy answers, who find the problems of a brewing crisis so intriguing that they cannot set them aside or forget them. Only those who really feel this way about research, and whose curiosity is never satisfied, are likely to make first-rate warning analysts.

b. Aptitude for detailed research. Unfortunately, not all the work involved in these basically fascinating and challenging problems is itself so fascinating. Like scientific experiments which involve a seemingly endless series of repetitive tests and much dull and monotonous work, much of the actual research of indications intelligence can be very tedious and time-consuming. No device has yet been worked out to spare the analyst from having to accumulate, sort and compile innumerable fragments of information. Some believe that in time computers may be able to take over a part of this and relieve the analyst of some of the tedium—but there will still be the problem of selecting the fragments to put into the computer system and making some type of assessment of them, which only the experienced analyst is likely to be able to do. Thus there is probably no way out, regardless of what tools to assist him may be devised, which will spare the analyst from spending much of his time in exhaustive (and often exhausting) research work. The good warning analyst thus must have more than an interest in the subject; he must also have an aptitude for, and willingness to undertake, all the detailed work which is involved.

It may be added that this attention to substantive detail is not only an attribute of good analysts; it is also an attribute of good supervisors even up to and including the chiefs of intelligence organizations. It has not been my experience that chiefs who delegate all substantive responsibilities and avoid reading the detailed basic material, or who regard their functions as purely administrative or supervisory, make good intelligence supervisors. The US intelligence system, at the intermediate and highest levels, has had both outstanding and mediocre chiefs. An indispensable attribute of the superior chief is that he has done his homework.

There is an old saying, usually applied to the hostess who would seek to give the flawless dinner party. It may also be the guideline, if not the achievable goal, of the indications analyst: "Details make perfection, and perfection is no detail."

c. Imagination. This point has been discussed, at least indirectly, in the preceding chapters. Basically, imagination is the attribute which permits the analyst to see beyond the obvious, to understand or at least attempt to understand the implications of unusual or extraordinary developments, and to be able to perceive the situation as others see it. It is an attribute of all creative thinkers, and intelligence analysts are no exceptions. While imagination alone, of course, does not make an exceptional analyst—and it is possible to bring too much imagination to intelligence problems and to lose sight of reality—it is nonetheless a characteristic of outstanding intelligence analysts. Few warning situations will be so clear-cut or so obvious that the perception of the "truth" will not require the exercise of the imagination as well as the compilation of the "facts." If the situation is that clear, there will likely be little need for analysis at all.

d. A retentive memory. Some studies have suggested that this may be one of the most important attributes for any kind of analytical work, and that the importance of memory as a factor in intellectual achievement has been greatly underestimated. Whether this is valid or not, there is no question that a good memory is invaluable to the warning analyst. The half-forgotten fragments set aside days or weeks ago may indeed be the missing clues. Since time (or rather the lack of it) is usually one of the greatest problems confronting the indications analyst, the ability to recall seeming trivia without having to make a search of the files (if indeed one has had time to do the filing) may make the difference between the good analytical effort and the superficial one, or even between the right answer and the wrong one. It has been demonstrated that motivation or interest often is the key element in memory—we remember that which is important to us, or that which interests us, and tend to forget or to ignore that which is unimportant to us. The outstanding warning analyst will often have a phenomenal memory for indications or potential indications which others have forgotten, but will not necessarily recall other facts or events which were not important as indications. Thus the discriminating memory—which is a product of interest and specialization as well as of an abstract ability to recall things generally without regard to their importance—is the prime requisite for indications analysis.

e. The recognition of that which is important. And, one might add, the ability to perceive that which is not important, or which is irrelevant to the solution of the problem at hand. There appears to be considerable variation among individuals in their ability to discriminate between that which is important and that which is not, even in their fields of specialization. A surprising number of analysts and supervisors seem to be readily distracted from the problem by information or developments which are either unimportant or largely irrelevant to the issue. Now, intelligence problems can be very complex, and warning problems are likely to be among the most complex of all. Analysts further are usually operating under considerable pressure and they are likely to be deluged with incoming material of the most widely varying degrees of credibility and importance. If they are unable to distinguish that which is of importance, or at least potential importance, from that which is not, they will be constantly led astray on wild goose chases and they will lose sight of that information which they should be considering. They are likely to be particularly vulnerable to sensational but inaccurate reports, and thus may spend an inordinate amount of time considering spectacular but largely irrelevant trivia while ignoring other basic intelligence of critical importance to the issue.

What causes this misdirection of effort, this inability to define the problem and isolate the material which is important to its solution and to set aside that which is not? Part of the difficulty, of course, may just lie in insufficient native intelligence. A second factor often will be insufficient education or training in the type of information or problem being considered—it is unfortunately not unheard of for judgments to be made on specialized substantive problems by persons who really are not qualified to do so, no matter how smart they may be. (It is often a bitter complaint of analysts that their expertise is ignored.) One cannot help feeling, however, that something more than these obvious factors is involved. Some people do seem deficient in that attribute known as "common sense" despite other intellectual abilities. In other cases, just plain fuzzy thinking or lack of logic seems to lie behind the inability to isolate the important data and to concentrate on it. In these circumstances, the perceptive supervisor or chairman of the discussion group or committee should seek to define the problem clearly and to require the analysts or the group to evaluate the relevance and significance of each piece of information to the problem. It is amazing what such an approach can sometimes do to focus attention on the important and relevant information and set aside the extraneous. A distinguishing characteristic of both good analysts and of good supervisors is the ability to do this.

f. The ability to entertain various hypotheses. The importance of considering alternative hypotheses, including unpopular or minority ones, was discussed in the preceding chapter. The ability to consider more than one solution to a problem unfortunately is not nearly so widespread among the college-educated as one would like to think. Such ability of course is closely linked to that of objectivity. In critical indications or warning situations, there are always people who seem unable to entertain more than one hypothesis or solution to the problem. They have already made up their minds on the likely course of action of the nation in question, and they evaluate (and even select) their data to conform with their hypothesis. Emotions and preconceptions are substituted for logic. (See discussion in Chapter 10.)

It is by no means easy, even for those who are making every effort to avoid emotional judgments and to maintain objectivity, to consider dispassionately two or more hypotheses. It is particularly difficult to do so in situations in which the facts are not provable or cannot be empirically demonstrated to be more relevant to one hypothesis than another—in short, where subjective as well as scientific criteria enter into the "solution." It is because this is so and that the facts may not "speak for themselves" that the consideration of various alternatives is both so important and so difficult for warning. No one should be misled into believing that it will be easy to achieve this. A deliberate effort must be made at every stage—from the analyst on up—to see that the unpopular hypothesis (or the one which may require the commander or policy maker to take some difficult action) is actually being given equal consideration. The analyst who is able to set forth dispassionately the case for both sides will be invaluable—provided, of course, he is also given an adequate opportunity to be heard.

g. Well-reasoned presentations. The analyst is more likely to be given that hearing if he or she is also capable of preparing a well-reasoned and logical analysis and of presenting it both in writing and orally. Brilliant analytical work will get nowhere unless it is disseminated in readable or bearable form to those who need to know it. It is not true, of course, that all people who do good research can present the results well. Although good researchers frequently write well, this is not always so. And many who write well are very inept at oral briefings, and would never be asked to speak twice. Training obviously will help; particularly on oral presentations, training in public speaking and practice in presenting briefings often pay dividends and will give the analyst an opportunity to be heard which he would otherwise lack.

The ability to present the evidence (or to see that someone else presents it) will be very important—and often will carry more weight than the most careful research which is poorly presented. The analyst who lacks the talent or training to present his case well should do something about it—either through his own efforts or by enlisting the help of his superiors. And supervisors should take care to ensure that good research efforts are not wasted for lack of someone to present the results.

h. Objectivity. We have addressed this before and we will do so again. Objectivity, objectivity and more objectivity should be the goal of every analyst, but above all of the indications analyst. It must also be the goal of his supervisor and of his supervisor's supervisor and on up the line. It may require a conscious effort at all levels to ensure that emotions and preconceptions do not prevail over reason.

Attributes of Character or Temperament

Important as the foregoing intellectual assets are, they will be insufficient if the analyst and his supervisor do not also possess certain important qualities of character and temperament.

a. Interest and motivation. The importance of these can hardly be overstated. These are the accompaniment of the insatiable intellectual curiosity which will ensure that it is applied to the problem at hand—obviously one cannot spread one's curiosity so thin that all problems are equally fascinating or there will be no time to apply oneself to the particular problem. In short, the successful warning analyst must not only be fascinated by intellectual puzzles in general, he must be specifically fascinated by warning and indications puzzles. It must be important to him, or to her. Innumerable studies have shown that interest, motivation and enthusiasm are more important to success than brains. In their field of specialization, good warning analysts are over-achievers.

b. An infinite capacity for hard work. No amount of intelligence or imagination or memory can replace the need to apply oneself to the problem. It's great to be smart, but it does not take the place of having studied the material, and studied it exhaustively. In warning problems, there is simply no substitute for having done the work. In crisis situations, sheer physical endurance and the ability to work long hours with insufficient sleep can be as important, perhaps even more important, than intellectual ability. This capacity for hard work, moreover, should not be limited to crises—when demands from on high and the great interest in the problem engender a sense of excitement which sustains enthusiasm. The analyst must also have a capacity for diligent application when it is not exciting and when in fact no one has asked for specific studies or shown much interest in his research. Obviously, willingness to work long hours in such circumstances is a product of high motivation rather than of any other type of incentive.

c. Initiative. For a variety of reasons—including the uniqueness and infrequency of warning problems and the relatively low interest in indications except in times of clear crisis—the warning analyst should have an exceptionally high degree of initiative. Unlike other fields of intelligence research, which have relatively set schedules of production or in which there is a constant stream of requests from above, indications analysts usually have to initiate almost all their own projects. In addition, they have to create a market for them and "sell them." In the twenty plus years which this writer has spent in the warning business, probably not more than ten percent of the production effort of indications analysts (except for a routine weekly watch report) has been requested by anybody. This handbook is no exception. Nearly all useful warning studies are initiated by analysts because they perceive a coming problem or crisis which they feel is not being adequately handled by the usual mechanisms of intelligence and which they feel needs to be given more attention. Moreover, they are quite likely to encounter a considerable degree of hostility to these independent projects or efforts to arouse attention or concern. Brick bats rather than praise for initiating the project are often the fate of the indications analyst. Needless to say, it requires an exceptional degree of initiative to persist in these unfavorable circumstances.

There is usually little reward for new ideas and most analysts will find that they get along best if they are not rocking the boat. And nothing is more likely to rock the boat than a suggestion that the intelligence community has missed coping with something important or is coming to the wrong conclusion about it or still worse, that the policy maker might soon be confronted with difficult and dangerous decisions.

d. Independence of judgment. This is a corollary of initiative and particularly of the willingness to assume the initiative when one is in the minority or the atmosphere is hostile. When it comes to judgments, the warning analyst will quite likely find himself a "loner." Indeed, only those who are temperamentally capable of making independent judgments—which may vary from those of the majority—are suited to be indications analysts. Those who find it easiest to go along with prevailing opinion or whose judgments are easily swayed because others (particularly their superiors) think differently have little to contribute to warning; they can be a positive handicap. Some analysts—even when assigned to indications offices which ostensibly exist to produce independent analyses—seem constitutionally unable to come to a different opinion from that of their superiors or their home offices or agencies. The "company man"—whose basic philosophy is "my agency, right or wrong"—no doubt has his place in the system. But it should not be in the warning portion of it. There is no point to having indications analysts who can only go along with majority opinion. It would be better not to have them at all.

To be able to make independent judgments, and to hold to them when they are criticized, the individual must have some confidence in his own ability and interpretations. Experience often will help, and so may the reading of history. The fact is, as we have noted in earlier chapters, that correct judgments in many warning situations (probably in most) have been reached by a minority of individuals. The well-reasoned independent judgment in such circumstances quite likely may be the right one.

e. Willingness to risk being wrong. There is probably no normal person who likes to admit that he has made an error. This is likely to be particularly true of intellectual judgments or opinions based on reasoning, since to admit that one has made a mistake in such circumstances also implies that one has been kind of stupid. It is different from betting wrong on the horses, or guessing wrong on a finesse; it is a matter of professional pride. As in all fields of intellectual endeavor, professional pride is important in intelligence. It is also particularly important to policy makers that intelligence be right. And it thus may be very difficult to move upward through channels an admission, either explicit or implicit, that an agency or an individual speaking for an agency has been wrong on a crucial issue. The results of this are two-fold: both the individual and the agency will seek to avoid making a firm judgment which might prove to be wrong; and once the wrong judgment has been made, it may prove extraordinarily difficult to reverse it. A preeminent former US policy official and senior advisor in intelligence has called "pride of previous position" one of the most dangerous pitfalls of intelligence.

The intelligence system needs to get over the idea—and the policy maker even more so—that intelligence should never be wrong, or, if it has been wrong, that it should never admit it. The analyst who fears to come to a judgment lest it be wrong or who can never admit that he made a wrong judgment contributes to the stagnation of intelligence. Not only will he tend to be so excessively cautious that he cannot commit himself at all, he may also postpone indefinitely the writing of items which might contradict anything he has prepared before. He may even, in his fear of admitting error, suppress evidence entirely which is contradictory to his previously expressed opinion.

The ability of the individual to admit error—or to take the risk of being wrong—is a product not only of moral fortitude but also of maturity and of confidence in one's own ability and judgments. Those who feel professionally or personally insecure (for whatever reason) are usually the most adamant in defense of their preconceived opinions and the most reluctant to admit that they could ever have made a mistake.

This is not something which the analyst should be expected to cope with alone, however. The intelligence system—up to the highest policy levels—should accept reality and recognize that intelligence cannot accomplish the impossible. It is not only ridiculous, but dangerous, for those in authority to expect perfection of intelligence. Some error is inevitable. It is better to have come to useful, positive judgments which are right most of the time, and to be wrong occasionally, than to be reluctant to come to positive judgments at all. In warning, above all, it can be fatal.

f. Indifference to rewards and appreciation. It is the fate of those who show many of the preceding attributes that they are likely to be bucking the system, fighting city hall, and often stepping on the toes of both their colleagues and superiors. And anyone who expects to receive much in the way of conventional awards or recognition in these circumstances should forget it.

The warning analyst who would do a good job should not expect also to be "popular"—indeed the better the work, the less popular he may be. Those who are continually challenging the facts or questioning the analysis or reasoning of others (particularly their superiors) are not likely to be highly appreciated—and it will not help in the least for them to be right. The individual who seeks to do a good job of indications analysis should accept at the start that his work will probably not receive much recognition and that he is unlikely to be the recipient of many awards. He may even—in particularly bitter controversies—have a falling out with his friends.

The warning analyst is likely to suffer from a constant sense of frustration. In routine times, most people are not interested in indications or warning and they do not wish to be bothered with problems that then seem remote. In times of crisis, the analyst has trouble getting people to listen, still more in persuading them to believe that he might know whereof he speaks. In either circumstance, he is likely to find that very few people will consult him and to feel that much of what he does is useless.

If there are nonetheless rewards in all this, they will probably be subjective rather than demonstrable—and to derive from the interest and fascination of the subject matter and a hope that in the crunch someone after all will listen, and that it will be important.

g. A sense of responsibility. Last, but by no means least, in this list of attributes of indications analysts one must put a sense of responsibility toward others. The analyst must feel—even he is wrong in this—that what he does matters and that the judgment of the intelligence system to which he contributes is indeed important to national security or to the safety of US or Allied forces.

If the analyst does not feel this sense of responsibility for the lives and welfare of others, if his approach to the problems before him is wholly academic, or if he feels that it is someone else's problem to "do something" about it—then he is more than likely to prove inadequate in the real crisis. It was a shattering emotional experience to find, in the summer of 1950, that there were analysts of the Chinese Communist military problem who seemingly saw no relation between their judgments on Chinese military deployments and the lives of US troops in Korea. They could not divest themselves of the idea that they had until next month or next year to come to some judgment on the destination and probable intentions of hundreds of thousands of Chinese troops. They apparently could really not understand that where these forces were was not an academic question and that their judgments of Chinese capabilities and intentions, if wrong, might be measured in the deaths of US servicemen. That was someone else's problem.

The analyst may indeed be a very little cog in the machine—and the bigger the machine gets and the more wheels there are, the less important any little cog may be. Nonetheless, the analyst must feel that what he does can make some difference, and that he does have some responsibility for national security. He cannot assume that someone else will take care of it.

In warning, you are your brother's keeper.

Can Warning Analysts Be Trained?

What a foolish question. Of course, analysts can be trained and are trained all the time. Why not as warning analysts too? Surely this presents no problem. Or does it?

Some interesting studies in recent years have begun to cast doubts on whether it is possible to retrain people, by the time they have achieved adulthood, in how to think. For example, a paper presented at the annual meeting of the American Association for the Advancement of Science in December 1968 by Dr. Robert B. Livingston suggested that the brain mold of the human being is largely set by the age of 12—not just psychologically, but physiologically—and that it is very difficult to make changes thereafter. Such an idea, if not the scientific demonstration of its validity, may not be so new: "Give me the child until he is five."

The new scientific inquiries into brain patterns and into how people, both as individuals and as members of groups or cultures, think and react under various situations may prove ultimately to be one of the most enlightening fields of inquiry for the understanding of intelligence analysis and the decision-making process. The writer at least feels that a better understanding of the work being done in the behavioral sciences in this area would be very beneficial to the intelligence system—probably more so than much of the effort now devoted to the computer field. It is also of course a highly controversial field; indeed it could be said to be dynamite.

The longer any person stays in the warning field, the more he or she usually comes to feel that how people think, how they reason, how they put facts together, and how much evidence they require before they are willing to come to judgments (or to change their previous judgments) is what the warning business is all about. This is what we need to understand better before we can decide what kind of training to give to indications analysts. What we need to know is whether there is a basic aptitude for this type of analysis—perhaps instilled at an early age—which really cannot be taught to the adult. Certainly, all of us know individuals whose approach to all controversial issues is so excessively conservative, or who are so reluctant to come to judgments before something can be positively demonstrated to be so, that they are of no use—in fact a positive hindrance—in the warning crisis. Nothing is likely to make perceptive warning analysts of people of this nature. They probably really cannot be made to change. This may not mean that they are misplaced in the intelligence field altogether—only that they are misplaced in warning intelligence.

Given a basic aptitude for this type of intelligence effort, and a positive motivation to do this type of work and do it well, however, the writer emphatically believes that there is much that warning analysts can learn, or can be taught. And so, hopefully, can their supervisors and the higher echelons of the government who are dependent on intelligence. There is much that can be learned from the experience of a live warning problem for those who are receptive to it. And, short of this, there is much that can be learned from history and the experience of others. And this is why this book is written.

13

Importance of Military Indications

For various reasons, some obvious and some less well-recognized, the collection and analysis of military data or indications is the predominant element in warning. By far the greater number of items on indicator lists deal with military, or military-related, activities. By far the greater portion of the collection effort, and particularly the most expensive collection, is devoted to obtaining data on the military strengths, capabilities and activities of enemy and potential enemy forces. This collection effort has greatly increased in the past decade, largely as a result of breakthroughs in technology, and the future promises a still greater expansion in the volume of military data. Whether this will necessarily result in a great improvement in our so-called "warning capabilities" is uncertain, although improvement in the timeliness of our evidence appears unquestionable. What does appear certain is that military information will consume even more of the intelligence effort in the future—and hence that military indications may assume an even more important role, or at least will take more of analysts' time, than in the past.

Primary Reasons for Importance of Military Indicators

At the risk perhaps of belaboring the obvious, it may be well to note some of the reasons why the intelligence community is so preoccupied with military indications and why they are so important. (In a later chapter, we will deal with the other side of the coin—why political indications are so important.)

First, and most obvious, military preparations are a necessity for war. There would be no warning problem, in the sense we use it in warning intelligence, if it were not that nations have armed forces and arsenals of modern weapons which they could commit against us or our allies. The nations of the world which are not thus prepared, whatever other problems we may have with them, are of little or no concern to us for strategic warning. The index of our concern about the intentions of other nations is largely how much military damage they can do to us, rather than just how politically hostile they may be, and our collection effort is usually allocated accordingly.

Secondly, many military preparations, although far from all, are physically discernible or at least potentially discernible to us. They involve movements of troops and weapons, or augmentations of them, which we can observe provided our collection is adequate. As a general rule (at least this has been true historically), the greater and more ominous the scale of military preparations, the more discernible they have been. The advent of weapons of mass destruction and long-range delivery systems, of course, has potentially somewhat altered this. It is now theoretically possible for a nation so equipped to make almost no physically discernible preparations for a devastating nuclear strike. Few analysts now believe, however, that there would not also be other potentially discernible military and political—indications before any nation undertook so terrible and perilous a course of action. The concept of a totally surprise nuclear attack (that is, the attack out of the blue without any prior deterioration in relations or military preparations of various types) no longer enjoys much credence. The problem of providing warning that the situation had deteriorated to the point that nuclear attack might be imminent is, of course, another problem. The effect of the advent of the nuclear age, in any case, has hardly been to reduce the importance of military indications! The result has been quite the contrary the community has devoted far more effort to attempting to determine what ancillary military indications might be discernible and to devising methods to collect such information.

One lesson which we have learned from crises of recent years is that discernible military preparations—the type of preparations which have traditionally preceded the outbreak of hostilities—have by no means lost their validity as indicators. The effect of such military episodes as the Chinese military intervention in Korea and the Soviet military intervention in both Hungary and Czechoslovakia has been to reinforce confidence in the value of military indications as less ambiguous and probably more dependable gauges of impending action than political, indicators. The amount of preparation undertaken by the Soviet Union and its allies for the invasion of Czechoslovakia has probably served to reduce fears—once quite prevalent—that the USSR from a standing start and without a discernible change in its military posture would launch a devastating attack against the West. That portion of intelligence sometimes defined as "hard military evidence" has gained stature for warning.

Another factor which lends importance or credence to military indications is that so many of them are so expensive to undertake—and that this is becoming increasingly so. It is one thing to undertake a relatively inexpensive propaganda campaign of bombast and threat, or to alert forces for possible movement or even deploy several regiments or so. It is quite another to call up a half million reservists for extended active duty, or to move a number of major ground force units hundreds or thousands of miles, or to initiate a crash program of production of new combat aircraft or naval landing craft.

Such serious, expensive, and often disruptive military preparations—and particularly those which take a real bite out of the taxpayer's income or which involve a commitment of national resources from civilian to military effort at the expense of the consumer—these are measures of how a nation or at least its leadership really feels about a problem and how important it is to it. These are the hard indications of national priorities. They are rarely undertaken lightly, or just for political effect or as a "show of force." There are cheaper ways to bluff or to make idle threats one has no intention of carrying out.

This is not to say, of course, that such major military preparations or reallocations of national resources are necessarily unequivocal indications of preparations for aggression. They may be defensive, or a recognition that the international situation is deteriorating to a point that the nation against its will may become involved in conflict. But such preparations are real and meaningful and important—a kind of barometer of what a nation will do and what it will fight for. And they are evidence that important decisions have been taken.

Such major changes in military allocations or posture or priorities are not only substantial or "concrete" indications; they may also be particularly valuable as long-term indications. They often give us lead time—tine to readjust our own priorities and preparations, not only in intelligence collection and analysis, but more importantly our own military preparations and allocations of resources. Provided we have recognized and understood them correctly, they may prevent our being strategically surprised even though our short-term or tactical warning may fail us.

Understanding the Basics How a Nation Goes to War

Know your enemy. This basic tenet is nowhere more applicable in intelligence than to the problem of strategic military warning. The analyst who would hope to understand what his adversary is up to in time of crisis should begin his education with the study of all that he can possibly find on the subject of how that adversary will prepare his forces for war.

This principle underlies the preparation of indicator lists, as discussed in Chapter 7. The well-prepared and well-researched indicator list should incorporate not only theoretical or general ideas of how the enemy nation will get ready for war; it should also include, insofar as is practicable, some specifics of what we know of the potential enemy's doctrine, practice and plans. As our knowledge of these things increases, we are likely to prepare better and hopefully more usable indicator lists.

It would be virtually impossible, however, to include in any indicator list everything *which* might occur if a major nation were preparing its forces for war. Where several nations might be involved (e.g., the Warsaw Pact countries), a list which attempted to incorporate all that we know, or think we know, about war plans, missions of specific units, wartime organization and terminology, civil defense preparations and any number of other subjects would become too cumbersome to cope with. Except possibly for very specific problems or areas, indicator lists are likely to remain fairly generalized. The warning analyst needs far more than this.

Other things being equal, the individual best qualified to recognize that a nation has begun serious preparations for possible hostilities should be the analyst who best understands its military doctrine and is best read in its military theory and practice. One unfortunate consequence of the separation of basic and current intelligence (when they are separated) is that the analyst who must make quick judgments on the significance of current information may not be well-grounded or up-to-date in such basic material (see that portion of Chapter 3 entitled "Warning is not current intelligence").

The distinguishing characteristic of preparations for hostilities (versus preparations for exercises or other relatively normal peacetime activities) is that they are *real*—and that they will therefore include activities rarely if ever observed in peacetime. Various exercises or mobilization drills or the like no doubt have rehearsed part of the plan, often in miniature form as a command post exercise, but almost never will the enemy have rehearsed in full force all the preparations that he will undertake when he is actually preparing to commit his forces. (Those who doubt this should reexamine the preparations for the invasion of Czechoslovakia.) An understanding of his doctrine and military theory, of what he did the last time that he was involved in a real combat situation and of what changes there have been in his military practice since then will be invaluable, indeed indispensable, to an understanding of what he is up to now.

The military forces of all nations are slaves in large degree to their doctrine and theory. What the staff officer has been taught in school that he should do in a given situation is likely what he will do. If doctrine calls for the employment of airborne troops in a given tactical situation, the chances are they will be employed. If the unit mobilization plan calls for the requisitioning of trucks from a local economic enterprise, the chances are very high that they will be requisitioned before that unit is committed to combat. If contingency wartime legislation provides for the establishment of a supreme military council (or some such super body), or for the redesignation of military districts as field armies, the warning analyst and the community should instantly recognize the significance of such developments should they occur. If the national military service law (normally unclassified) provides that reservists should be recalled for no more than three months in peacetime, evidence that they are being held longer could indicate that emergency secret wartime decrees had been passed. And so forth. Innumerable and much less obvious examples could no doubt be cited by experts on such topics.

Unfortunately, the experts on such topics are often scattered and engaged in long-term basic research, even when a crisis is impending. Means must be established so that this information is not overlooked or forgotten, or filed in the library, when it is most needed. (One possible method for coping with this problem was suggested in Chapter 9.)

How many analysts assigned to the current or indications effort at various headquarters or various echelons in the national intelligence community have read or studied much of the basic material that might be so crucial to the understanding of actual preparations for hostilities? How many have if catalogued or readily available or know where they could immediately obtain the answers they might need? Some offices and analysts no doubt are much better prepared than others, but experience suggests that no intelligence office is really fully prepared now for the contingency of real preparations for hostilities by one of our major potential enemies. Every major crisis, at least, has shown this to be true. And those at the supervisory or policy levels, for obvious reasons, are quite likely to have minimal knowledge of a great deal of basic information which could be crucial in a real crisis.

Warning specialists and current military indications analysts could make no greater contribution over the long term than to do their best to review, study and compile the rarely reported and often obscure basic data which might some day be so essential for warning. The scope of potentially relevant military information is vast: doctrine, theory, logistics deficiencies and wartime requirements, applicable legislation, mobilization theory and practice, military terminology, major military exercises and war games, combat readiness and alert procedures, and other similar basic topics. The study of the performance of our potential enemies in relatively recent live crises, as we have noted before, is also invaluable even though the situation the next time will doubtless not be identical. It is not even too much to suggest that there are still many useful lessons in warning to be derived from a study of World War II. The German deception. effort prior to the Battle of the Bulge and that of the Allies prior to the Normandy invasion are classics—and live reminders of how effective deception can be (this topic will be discussed in Chapter 29: "Deception: Can We Cope with It?") The Soviet buildup for the attack on Japanese forces in Manchuria in August 1945, is a useful example of the difference in strategic and tactical warning. And the campaign itself, as well as earlier operations of Soviet forces in the same area, provides potentially useful lessons in how the USSR might go about conducting operations against Chinese Communist forces today.

14

Order-of-Battle Analysis in Crisis Situations

The best-known, most venerable and most prestigious field of military analysis is "order-of-battle." In its broadest sense, it is sometimes used to mean almost the whole scope of activity of military forces. In its narrower (and more correct) usage it is defined (by the *JCS Dictionary*) as: "The identification, strength, command structure, and disposition of the personnel, units, and equipment of any military force." It is in this sense that it is used in this chapter and throughout this handbook.

It is readily apparent that a determination of the order-of-battle of enemy forces is of decisive importance for warning intelligence. Indeed, insofar as warning rests on a determination of the *facts*—as opposed to the more complex problem of determining what the facts mean and issuing some interpretive judgment the order-of-battle facts will often be the single most important element in warning. Whatever other facts may be relevant or significant for warning, nothing is likely to be so critical as the locations, strengths and equipment of the enemy's military forces—for these determine what the enemy *can* do. And, as discussed in Chapter 5, the understanding or correct assessment of the enemy's capability is a prerequisite to the assessment of intentions, and the failure to recognize the capability is fraught with peril. Thus, obviously, at every step in the process, order-of-battle analysis will play a crucial role in the issuance of warning. And order-of-battle analysts will carry a heavy responsibility.

Order-of-Battle Methodology

Since most nations are usually at peace and the strengths and locations of their military forces are relatively static, or at least the changes are fairly gradual, order-of-battle analysis normally tends to be a rather slow and hence conservative process. Over a period of time, certain methodologies or criteria are established (again rather conservative ones) which determine, by and large, whether a given unit can be "accepted" in the order-of-battle. While the methodologies will vary somewhat from country to country and among various types of forces (i.e., the same criteria for "acceptance" may not apply to a Soviet ground force unit as to a North Korean air unit, still less to a Vietnamese guerrilla unit), the criteria nearly always are relatively rigid. In particular, they call for a relatively high degree of substantiation or "proof" that the unit does in fact exist, is of a certain strength or echelon, and is located in some specific area. Also, some unit number (hopefully the correct one, but at any rate *some* identifying number or designation) is highly desired by OB analysts. A unit which must be described as "unidentified" (u/i) lacks status, as it were, and is a little suspect. There is always a possibility that it is already carried elsewhere in the OB as an identified unit and has moved without detection. The analyst always seeks to "tidy up" his order-of-battle so as not to have such loose ends. Ideally, all units are firmly identified and located on a current basis, together with their correct designations, commanders and equipment, and their manning levels are consistent with the accepted table of organization and equipment (TO&E) for that type of unit. Everything fits. No problems. All units have been "accepted" and we have confidence in the OB.

Now, this ideal OB situation, of course, is rarely met—and almost never so in hostile nations which are attempting to conceal such military facts from us. For a number of reasons—the difficulties and slowness of collection, the conservative criteria for acceptance of units, the personal views of analysts, the requirements for inter-agency or sometimes international consultation, and even the slowness of the reporting, editorial and printing process—the accepted order-of-battle is almost invariably out-of-date. And for some units, it may be years out-of-date. Order-of-battle analysts are normally loath to admit it—and few outside their immediate circle probably realize it—but it has by no means been unusual to have delays of two and three years in the acceptance of new units, determination that units have been inactivated, or recognition that they have been upgraded or downgraded in strength or relocated, redesignated, resubordinated, converted to another type, or split into two or more units. Incorrect numerical designations of units have been carried in OB summaries year after year; only by chance will it sometimes be found that units have been incorrectly identified for five or even ten or more years. Quite often, the last thing we can learn about a unit is its numerical designation.

In short, the "accepted" order-of-battle and the real current order-of-battle for enemy nations probably never are identical. The most we can hope for is that it is reasonably close to the facts and that we have made no serious misjudgments of the enemy's strengths and capabilities. In time, we will usually obtain the data which we need to correct the OB—and we can change the location, designation or whatnot on the next printing. It is better to be slow or late with a change than to make a judgment too soon that may be erroneous and which we will have to retract later. Such, in general, is the philosophy of OB analysis. This is not intended as criticism; it is a statement of fact.

And most of the time, it does not really matter that it is like this—which is probably the chief reason that this normal lag on OB acceptance is so little recognized. No one suffers from errors in the order-of-battle or from delays in information that a unit has been formed, upgraded or moved. Even egregious errors in estimating the strength of foreign forces (or particular components of foreign forces—like the great overestimate of Soviet missile strength in the late 1950's) have not caused us any real harm. Some might even say that the effects for budgetary purposes have been beneficial. Few care and no lives have been lost.

Except when there is a warning problem when we or our allies are in danger of being attacked. Then these order-of-battle "details" can matter and matter decisively. When enemy units are being upgraded or moved or mobilized or otherwise prepared for combat, then the accuracy and currency of order-of-battle *do* matter and lives can be lost—and many have been—for errors.

Often, however, sudden changes in the enemy's military situation—including a redeployment of units for possible commitment to combat—have not brought forth imaginative and responsive changes in methodology by order-of-battle analysts. It is in fact the usual case that the criteria for "acceptance" of changes in unit strengths and locations are not modified—even when it is clear that major troop movements are in fact in progress.

Sudden changes in unit locations, particularly large and secret redeployments, admittedly pose tremendous problems for order-of-battle analysts. It is rare indeed that the initial evidence will be so good, or accurate or complete that it will be possible to determine with any confidence how many troops are involved, what types or how many units are deploying, still less their unit designations or where they come from. Time is required to sort out the data, to attempt to obtain coverage of the home stations of the more likely units to see if they have in fact departed, and it will probably be weeks or months (even with good collection) before some of the needed information can be obtained. It may even be years—or never. To this day, we have only partially reconstructed the order-of-battle of the Soviet units which intervened to suppress the Hungarian revolt in November 1956, and it is likely that we will never know the total strength and units involved.

Confronted with initial reports—which may include some highly reliable observations and other good evidence as well as hearsay and unconfirmed reports that large numbers of troops of unknown designation are moving from unspecified locations toward undetermined destinations, the reaction of the normal OB analyst not surprisingly is to wait. What units can be moved on his situation map? He has no identifications, no firm evidence precisely what area the troops have come from; he may not be sure whether one or ten divisions could be involved although he usually suspects the lower figure and that many of his sources are greatly exaggerating the facts. The whole situation is anathema to him; it would be in violation of all his criteria to move any unit until he knows *which* units to move. It may be all right for current or indications analysts to talk vaguely about large but unidentified troop movements, but the OB analyst must be specific and precise. He must "accept" or decide not to accept the movement of specific units (and this might well include new units of whose existence he has yet to learn). Confronted with this dilemma, he moves nothing—not yet.

The Indications Approach to Order of Battle

On the other hand is the indications analyst, sometimes but not always supported by other current analysts, who cares less for the order-of-battle details than that the community and the policy maker recognize, and recognize *now*, that the enemy is deploying major forces and that there is grave danger that this buildup is preparatory to their commitment. He implores his colleagues and superiors to see what is happening—and he is driven half frantic when they turn to the order-of-battle "experts" whose reply is that they "cannot accept" that yet. And the indications or current analyst *may* even be told that he cannot report a *possible* increasing threat—how can the threat be increasing when the OB map does not show any buildup of forces in that area? Perhaps it does not even show any units there.

This is not a hypothetical fiction or an indications analyst's nightmare. This is what can happen unless an impartial arbiter in position of authority intervenes to hear both sides. It is unquestionably true that the rigid criteria of order-of-battle acceptance have held back warning for weeks, and that this can be the most serious single impediment to the issuance of military warning. A few examples will illustrate the point more convincingly.

Not long after US and UN forces intervened in the Korean conflict in June 1950, reports began to be received of northward troop movements in Communist China. Many of these reports indicated only that large numbers of troops were leaving the southernmost provinces of China (there had been a substantial buildup in this area to complete the conquest of the mainland the preceding year). By late July, it was clear that substantial elements of the Fourth Field Army had left South China. Meanwhile, there were numerous although somewhat conflicting reports concerning troop movements farther north—while there were many reports which indicated that troops were, moving to Manchuria, it could not be confirmed how many of the troops were proceeding that far and how many might have deployed to intermediate locations—including Fukien Province for a possible assault on Taiwan. Even as reports continued to mount during August of heavy northward troop movement (some reports from Hankow claimed that troop trains were moving northward day and night), there was little evidence on the whereabouts of *specific* units or sufficient data to make a reliable estimate of the number of troops involved. Indications and order-of-battle analysts locked horns over the issue—with warning analysts pleading that the community recognize the likelihood of a major buildup in Manchuria, while OB claimed that there was "insufficient evidence" to move *any* units to Manchuria. Not until the end of August were watch reports (reviewed by OB) permitted to go forth stating that any elements of the Fourth Field Army had moved to Manchuria (attempts to say that "major elements" could be there were vetoed). Not until about 1 October did order-of-battle "accept" that elements of some six armies of the Fourth Field Army were in Manchuria—and make its first increase in estimated strength in that area. The criteria for OB "acceptance" actually held back warning of the buildup in Manchuria for at least six weeks, and this discrepancy as to what could be reported or "accepted" as evidence continued to a lesser degree up to and after the major Chinese Communist offensive in late November. Who can say whether or not an earlier acceptance of the buildup would have increased the likelihood that US and UN forces would have been better prepared for the Chinese onslaught?

A simpler and less damaging example of the difference between the order-of-battle and indications approach to troop movements may be derived from the Hungarian revolt in 1956. The Soviet response to this unexpected explosion was to pour troops into Hungary over virtually every road and rail connection from adjacent areas of the USSR. There was no reliable information on the total forces involved in the buildup; it was only clear that there were a lot. Concurrent with this heavy troop movement from the Carpathiari Military District, an extraordinary travel ban was imposed in Romania, prohibiting all attaché travel to rail centers north of Bucharest—while several informants reported to the US Embassy that Soviet troop trains were moving northwestward through Romania during the same period. By the criteria of indications analysts—for warning purposes—this was sufficient evidence to justify a statement that Soviet troops from the Odessa Military District (from unknown units and in unknown strength) were *probably* deploying to Hungary. To order-of-battle analysts, such a conclusion could not be justified. They stated flatly that they could not accept information of this kind. About six months later, a unit number associated with a Soviet division from the Odessa Military District was identified in Hungary, and the division was immediately accepted in the order of battle.

By the time that Soviet forces were deploying for the invasion of Czechoslovakia in the summer of 1968, our collection and analysis had seemingly improved, or perhaps because we had more time we were able to do better. At any rate, the usual conflict between indications and order-of-battle analysts was subdued, and there were fewer (one cannot say no) complaints from warning analysts that too few units were being accepted by order-of-battle or accepted too late. A controversy did arise, however, over the Soviet troop movements into Poland which began in the last few days of July—which illustrates that this question of troop movements when unit identifications are not available remains a potentially serious problem for warning. The initial sources of information on these Soviet troop movements into Poland (later determined to total five divisions from the Baltic Military District and probably two divisions from areas farther south) were primarily tourists and other legal travelers in Poland. Within 48 to 72 hours after the initial movements into Poland began, the US Embassy in Warsaw had received a half dozen voluntary reports from travelers of sightings of heavy troop movements at several road crossing points from the USSR. Within a few days, Western attaches had located two holding or assembly areas for these forces—one to the north of Warsaw and one to the south—to which they were denied access. Indications analysts, and to a lesser degree other current analysts, were prepared to "accept" or at least to report that substantial Soviet forces were entering Poland which were apparently backup forces (not part of the concurrent buildup directly along the Czechoslovak border). Even indications analysts, however (usually considered alarmist or prone to exaggerate by their more conservative colleagues), would hardly have been willing to accept in early August what later proved to be true—that some 11 or 12 Soviet divisions all told were then in Poland, as compared with the normal two. Order-of-battle analysts, using their criteria, were not willing to accept any units which they could not identify, or whose movement from home stations had not been confirmed, which resulted in some delay in acceptance of additional Soviet divisions in Poland.

Concealment and Detection of Various Types of Movements

A few comments may be useful here concerning our capabilities to detect changes in the status of various types of units and varying types of movement.

The foregoing discussion has discussed ground force units exclusively, for two reasons. In most military buildups, the great bulk of the deployments will be ground force units and their status and capabilities will usually be the major concern. And secondly, ground force movements are usually the most easy to conceal and hence the most difficult to detect on a timely basis, so that the major order-of-battle problems for indications and warning are usually, although not always, with respect to the movements of ground force units.

It should be fairly obvious why this is so. Ground force units, although they may be the largest in numbers of men and total piece of equipment, can be broken up for deployment and moved by a variety of means in small contingents. The men and much of the equipment often can be rather easily concealed or camouflaged. There is a great deal of territory in which units can be deployed, and it is very difficult to cover it all. Even very fine and frequent aerial reconnaissance (except in exceedingly open or desert terrain) may fail to discover many troops or pieces of equipment—and less adequate collection means may mean that major formations and movements are wholly undetected.

Aircraft and naval units, on the other hand, are much more difficult to hide, even though, there may be many fewer of them, and their movements usually are more likely to be detected. Aircraft can only operate from airfields, or from improvised air strips of some type, in open terrain. While the aircraft themselves may be partially or even wholly concealed in hangars or caves, major aircraft deployments require the use of major and fixed ground facilities. It is very difficult indeed, if some collection assets are available, totally to conceal the redeployment of hundreds of fighter or bomber aircraft. Moreover, it is much easier to count them and hence to come up with an accurate order-of-battle than it is for ground force units.

Naval units have the advantage, at least sometimes, of vast oceans in which to move. But, except for submarines or in cases of extremely poor weather, naval units cannot be concealed from observation on the open seas. Once found by sea or air reconnaissance, naval movements are extremely vulnerable. And, like aircraft, they are very easy to count and identify when they have been found.

This is not to suggest that there have not been major—and extremely successful—instances of secret deployments of major air and naval forces, as Pearl Harbor should remind us. The problems of order-of-battle analysis, and the arguments between OB and warning analysts, apply to all types of forces in some degree. It is just that, in most cases, they are more likely to be acute for the ground forces, where the order-of-battle lag is usually the longest.

The method by which troops and equipment are transported may be a significant factor in the timeliness of our detection. Generally, in relatively denied areas, the most difficult deployments to detect are those made by rail, since military equipment and freight are often not readily distinguishable and relatively few people will be aware of major changes in rail schedules or movements. Deployments by road are easily observed, are likely to disrupt traffic and usually will be common knowledge among the local citizenry; the chances of some reporting of such movements are usually better than by rail. When aircraft or ships are used to deploy troops, we are the most likely to detect the movement of the transports—but, unless coverage is very good, we may not be able to tell whether troops, civilians or cargo are being moved. We may be alerted to an abnormality as in the case of the Soviet ship movements to Cuba in 1962—without being able to determine initially what is being moved.

Needed: A Voice for Warning in the OB Buildup

The differences in approach, in criteria, in analytical techniques between indications and order-of-battle analysts are serious, and they are potentially as damaging for warning in the future as they have been in the past. Indeed, a dependence on the strict criteria of order-of-battle in a genuine crisis in which the enemy was employing his most sophisticated security and deception techniques could be absolutely catastrophic. At the same time, few would presume to suggest that the normal analytic techniques of order-of-battle (which have generally worked well, if slowly, in peacetime) should be tossed overboard for more imaginative, less precise and more "indications-oriented" techniques.

What is the answer to this serious dilemma? The answer is that both sides should be given their say, but that order-of-battle analysts should not be permitted in the crisis situation to have the last word on what can be "accepted" or what can even be reported to higher authority. There are things more important in an impending showdown with the enemy than the purity of order-of-battle techniques. It is more important that superior authorities know that some enemy troop buildup is under way, even if we cannot be too precise about it, than that they be led to believe that there is no such buildup.

The arbitration of such disputes belongs at the supervisory level. The system must not work—as it has sometimes worked in the past—so that the order-of-battle analyst is permitted to review and to veto what indications and current analysts, in their best judgment, believe to be of warning significance. For the most part, order-of-battle analysts are accustomed to a degree of autonomy and independence on what they will decide or "accept" which is probably unparalleled at the analytical level of intelligence. Military analysts on other subjects, and political and current analysts in general, do not begin to have the authority to make judgments which the order-of-battle offices take for granted as their prerogative. Nearly everyone—regardless of their experience in political analysis—feels free to question the judgments of political experts, or to debate the meaning of propaganda statements. Much military information is freely discussed and analyzed and varying opinions or judgments as to its significance may be offered, often by those who make no claim to particular expertise on the subject. But for some reason order-of-battle analysis (together with a few other relatively technical or specialized subjects) has usually been considered almost solely the responsibility of order-of-battle experts—and as a general rule they, and they alone, have been permitted to make the assessments as to whether or not a given military unit does or does not exist or has or has not deployed. And often the basis or reasons on which such judgments are made, or not made, goes unquestioned by higher authority. Intelligence chiefs and policy officials may well not even hear contrary argument or opinion on troop deployments or other order-of-battle material unless the indications or other current analysts are given an equal opportunity to present their views.

To paraphrase a famous quotation, "War is too important to be left to order-of-battle analysts." With all due respect to their experience and expertise, it is really expecting too much that they will change their techniques, methodology and criteria for acceptance or reporting of units when some extraordinary situation arises—and when it is imperative that judgments be made soon lest the warning be too late.

What the intelligence community needs in these circumstances is to recognize—and give equal time to—both order-of-battle and warning analysts. It needs on the one hand the analysis and judgments of order-of-battle experts—who will generally be adhering to their traditional, and conservative, techniques. It will be important that the supervisory level understand (and this may mean to inquire actively into) these techniques and criteria, so that it will be appreciated how much firm or unequivocal evidence is actually required by order-of-battle before a unit is accepted. The supervisor should have some appreciation of the normal time lags in identification and acceptance of units so he may recognize how long it might be before confirmatory order-of-battle evidence may be received. Subject to such supervision, the "confirmed" or "accepted" order-of-battle would continue to be reported much as it would in normal circumstances.

In addition, however, there should be another level or type of reporting which would concern itself with the indications of what *may be* or *could be* under way. Since the order-of-battle analysts often will be reluctant to make any judgments on information they cannot "accept," it may be imperative for warning that another group of analysts, not bound by order-of-battle criteria, be given equal opportunity to report the indications and to make at least tentative assessments of their significance and implications. It is essential that this reporting of what is possible not be subject to the veto of an order-of-battle analyst because it does not meet his traditional criteria for acceptance.

Conventional order-of-battle methods must not be permitted to hold back warning. To ensure that the indications of a military buildup are being adequately reported, and not just that which has been "accepted," is a constant and most important responsibility of the supervisor. He must not attempt to resolve the argument by turning it back to the order-of-battle analysts for decision on the grounds that they are the "experts." To do so may be fatal to the warning judgment. Over a period of years, this issue has probably been the single greatest bone of contention between the indications system and the rest of the community. To deny the warning analyst an equal hearing on the all-important issue of the military buildup is to make a mockery of the whole indications system—and to defeat the purpose for which it was established. The so-called "independent" indications effort will have no status whatever if it is not accorded a voice on this crucial issue.

It would be misleading to close this chapter without noting that progress has been made on this problem. Over a period of years many people have come to recognize that normal order-of-battle methods may be inadequate and too slow in crisis situations. Post-mortems have often confirmed the contentions of indications analysts that their methods, in these circumstances, are more accurate and responsive to the problem than waiting for all the order-of-battle "proof" to come in. The proliferation of indications shops and general dispersion of the military analytical effort in the community, although probably excessively duplicative in some respects, also has done much to ensure that order-of-battle analysts do not have a monopoly on reporting military movements, and their techniques have come under increasing scrutiny by other analysts. Nonetheless, this is a problem to which the community at all levels needs to be constantly alert. Particularly when an abnormal situation is developing, and it becomes evident that unusual troop movement is or could be in progress, the intelligence system may need to take specific steps to ensure that indications of order-of-battle changes, as well as confirmed deployments of known units, are being adequately analyzed and reported.

15

Analysis of Mobilization

TRADITIONALLY IN EXPECTATION OF WAR, or sometimes not until after it has broken out, nations great and small have mobilized. Probably all developed nations, and most under-developed ones as well, have some type of national or general mobilization plans. Historically, the declaration of general or national mobilization often was the decisive indication of the imminence or inevitability of war—a sign that the die was finally cast. So it was on the eve of World War I:

> In Berlin on August 1 [1914], the crowds milling in the streets and massed in thousands in front of the palace were tense and heavy with anxiety. . . . The hour of the ultimatum [to Russia] passed. A journalist in the crowd felt the air 'electric with rumor. People told each other Russia had asked for an extension of time. The Bourse writhed in panic. The afternoon passed in almost insufferable anxiety. . . . At five o'clock a policeman appeared at the palace gate and announced mobilization to the crowd, which obediently struck up the national hymn, 'Now thank we all our God.' Cars raced down Under den Linden with officers standing up in them, waving handkerchiefs and shouting, 'Mobilization!'
>
> Once the mobilization button was pushed, the whole vast machinery for calling up, equipping, and transporting two million men began turning automatically. Reservists went to their designated depots, were issued uniforms, equipment, and arms, formed into companies and companies into battalions, were joined by cavalry, cyclists, artillery, medical units, cook wagons, blacksmith wagons, even postal wagons, moved according to prepared railway timetables to concentration points near the frontier where they would be formed into divisions, divisions into corps, and corps into armies ready to

> advance and fight. One army corps alone—out of the total of 40 in the German forces—required 170 railway cars for officers, 965 for infantry, 2,960 for cavalry, 1,915 for artillery and supply wagons, 6,010 in all, grouped in 140 trains and an equal number again for their supplies. From the moment the order was given, everything was to move at fixed times according to a schedule precise down to the number of train axles that would pass over a given bridge within a given time.[1]

Because of the criticality of the initiation of general mobilization, its almost certain impact on or warning to the enemy, and because of the enormous commitment of national resources involved in a general mobilization, nations have sometimes deferred national mobilization until after war has broken out. Thus, both Britain and France declared full or general mobilization only after Hitler's invasion of Poland on 1 September 1939, although both powers almost certainly regarded war as imminent and inevitable following the announcement on 23 August of the conclusion of the German-Soviet non-aggression pact, and both nations of course had taken a number of important preparatory steps and partial mobilization measures prior to 1 September. Nonetheless, so long as the faintest hope remained that Hitler might refrain from attacking Poland and probably also to ensure popular support, the leadership of both countries deferred the crucial step—so long as it could be avoided.

Needless to say, the national mobilization measures which preceded or accompanied the outbreak of both world wars were scarcely secret—nor was any particular effort made to keep the effort secret even on the part of the aggressor, Germany. The coming of both wars was clearly apparent, and accompanied even by such conventional diplomatic moves as the issuance of ultimatums, and there was no need whatever to attempt to conceal national mobilization—which in any case would have been impossible.

The intervening years have in no way diminished the importance of mobilization as an indication of hostilities. While public announcements of mobilization may no longer be expected from nations seeking to conceal their intentions, a full national mobilization probably remains the single most valid indication of war. Nor is it true that our potential enemies today pay less attention to mobilization because of the possibility that the war could last only a few hours or days in a great nuclear exchange. Some people on our side may contend that modern war will make mobilization obsolete, but the Communists have yet to learn this. On the contrary, they appear to have placed even greater emphasis on detailed mobilization plans and their meticulous implementation than was true in the past. Units have explicit mobilization plans, most reservists know precisely where to report and all aspects of the mobilization plan receive the most careful attention and review. One obvious reason for this—apart from the fact that war will not inevitably be nuclear—is that the possibility of nuclear war greatly reduces the time in which mobilization may have to be implemented. National survival may virtually require that the whole process go into effect the moment that war appears imminent—or when the military units themselves are being brought to the highest stages of combat readiness. To delay could upset the entire plan for both defense and offense. Thus, if anything, a major mobilization today appears more rather than less likely. But chances also are that our enemies would seek to implement it very rapidly (we know that they are placing tremendous emphasis on speed of response to alerts and rapid transition to full readiness), so that our warning time might be very short.

Further, from a series of post-war crises we have learned that partial mobilizations are valid indications of preparations for limited hostilities or at least for the possibility of such hostilities. We have also learned that even forces which we have considered at very high levels of preparedness may require some mobilization at least of their support elements. One of the intelligence benefits of the Soviet invasion of Czechoslovakia (of which there were many) was the discovery of how much mobilization was needed or at least undertaken in support of this relatively low-risk operation. The result has probably been a growing appreciation of the importance of mobilization as an indication and also a better understanding of what we should be looking for.

Types and Categories of Mobilization

There are varying types or degrees of mobilization, which may include economic as well as purely military measures, and it may be well to define in general terms the various steps which may be involved.

A *partial* or *selective* mobilization normally will involve a recall of reservists to fill out certain under strength units or the activation of reserve units (such as National Guard units in the United States). Sometimes, although not always, this limited recall will be accompanied by a deferment of releases of men due to be discharged, and/or the callup of larger than normal numbers of recruits for training. The latter step usually is taken only if the potential need for additional manpower is deemed to be of some duration. If a relatively short-term crisis or requirement is foreseen, a recall of trained personnel or retention of trained men obviously is the only satisfactory expedient. Thus, the type of selective or partial mobilization chosen may provide some insight into how the nation in question views the crisis or problem. In any case, however, a partial mobilization usually is taken for one of three reasons: (a) as a contingency preparation when the enemy appears to be embarked on action which could result in conflict (e.g., the US recall of reservists in the Berlin crisis in the summer of 1961); (b) when a nation is preparing for a limited aggressive action of short-term duration (e.g., the Soviet invasion of Czechoslovakia); or (c) when a continuing requirement for additional manpower for the armed forces is anticipated for some time to come (e.g., the raising of US draft quotas for the Vietnam war, or the considerable expansion of Soviet forces beginning prior to the start of the Korean war and continuing into the fifties). Thus, a partial mobilization (usually, if not always) does involve an expectation or possibility of conflict—but not total conflict or a major commitment of the nation's resources.

Full mobilization, on the other hand, is rarely if ever undertaken except in expectation of major conflict of either short or long duration. A full mobilization of the armed forces will involve the filling out of all existing units, the activation of all reserve or cadre units, and the mobilization of all necessary support units and logistic backup required to bring the armed forces to full combat readiness. Needless to say, this a serious and expensive step for any nation, not be undertaken lightly, and historically it has nearly always meant that a major conflict was imminent or under way.

General mobilization probably would be considered the equivalent of full mobilization. It involves the activation of the nation's wartime plan for its armed forces—the recall to duty *of* all active reservists and the mobilization of all units in the approved wartime force structure. The day when full or general mobilization is put into effect is M-Day.

Total mobilization, while not always differentiated from *full* or general mobilization, is a useful descriptive term to define that mobilization above and beyond what the nation had initially planned for. It will involve the expansion of the armed forces beyond M-Day plans, the formation of totally new units, and the full requisitioning of the resources of the nation to sustain its armed forces. There is implicit in the term a sense that the nation is literally fighting for its survival. The USSR and Nazi Germany were totally mobilized in World War II. The US was not. A classic definition of total mobilization was the edict of Haile Selassie to his nation on the invasion of Ethiopia by Italy in 1935:

> Everyone will now be mobilized and all boys old enough to carry a spear will be sent to Addis Ababa. Married men will take their wives to carry food and cook. Those without wives will take any woman without a husband. Women with small babies need not go. The blind, those who cannot walk or for any reason cannot carry a spear are exempted. Anyone found at home after receipt of this order will be hanged.

Industrial or *economic mobilization* obviously involves the productive resources of the nation as distinguished from its armed forces. General mobilization probably, and total mobilization unquestionably, will also involve economic mobilization. In its extreme form it will involve the diversion of all economic resources, other than those needed for the bare sustenance of the civilian population, to support of the war effort. Both great industrial and primitive agricultural nations can be economically mobilized—both Germany in World War II and North Vietnam in recent years meet the definition. Since there is a later chapter on economic indicators, we will confine the discussion in this chapter to military mobilization.

Pitfalls of Mobilization Analysis

The intelligence system is more likely to obtain data on and to report promptly the movement of a single submarine or the redeployment of a squadron of aircraft than it is the callup to the armed forces of a half a million men. This extraordinary phenomenon—of which many people need to be convinced—is testimony to the enormous differences between democracies and dictatorships. Information which could never be concealed in this country, even in wartime, is routinely held secret in Communist nations and protected by highly effective security measures. Information known at least in part to thousands or even millions of people in the USSR or Communist China may be unknown to us. And among those secrets is likely to be the conscription quotas and strengths of the armed forces, including considerable variations in their strength and partial mobilizations in times of crisis. Compounding our problem may be excessive requirement for "proofs" of such changes which is almost unobtainable rather than a more flexible approach which would permit us simply to acknowledge that major increases seem to be in progress, the extent of which we cannot judge with certainty.

In a speech to the Supreme Soviet on 14 January 1960, Nikita Khrushchev announced a series of strength figures for the Soviet armed forces from 1927 to 1960 the first time certainly since World War II and possibly ever that the Soviets had released such data. Among the figures announced by Khrushchev for the total strength of the Soviet regular armed forces (exclusive of internal security troops) were:

1945	11,364,000	(presumably the World War II peak)
1948	2,874,000	(result of post-war demobilization)
1955	5,763,000	(increase attributed to US-NATO threat)
1960	3,623,000	(result of claimed reductions of 2,140,000)

Now the Soviet Union traditionally, and Khrushchev specifically, have not been noted for the reliability and completeness of their public statements, and it would be naive to assume that one should accept these figures precisely at their face value. It is a characteristic of published Soviet statistics that they are very sparse and, while usually accurate insofar as they go, are so incomplete that they can be very misleading (like the Soviet published military budget). Further, in the case of this particular speech, political objectives undoubtedly were the motivating reason for these pronouncements which also included a proposal (dutifully ratified by the Supreme Soviet) to reduce the Soviet armed forces by another 1,200,000 men. In addition, available data, lead us to believe that the peak of Soviet armed forces strength in the 1950's was not reached in 1955, as implied by Khrushchev, but a couple of years earlier in 1952–53 or specifically because of the Korean war.

Allowing for such discrepancies (which distorted the reasons for and probably the timing of the Soviet military buildup in the early 1950's), Khrushchev's figures probably reflected accurately the basic trends in and general magnitude of Soviet armed forces strength—or so we recognize now.

Most interesting was his claim that the strength of the armed forces had. more than doubled in the period from 1948 to 1955 (an alleged increase of 2,889,000 over the 1948 figure of 2,874,000). For indications analysts this was of particular interest since they had repeatedly urged an increase in the US estimates of Soviet strength during this period on the basis of evidence which seemed incontrovertible. Among that evidence had been the buildup of a number of cadre units (including two mechanized armies in East Germany) in 1949, a substantial increase in callups to the armed forces (including a two-class callup in 1951), and the active if unannounced participation of Soviet air and air defense forces in the Korean war.[2]

And what were the current estimates? An increase in Soviet air and naval forces was reflected in going estimates between 1950 and 1955—the estimated strength of the air forces rose from 650,000 to 835,000 and that of the navy from 600,000 to 695,000 (US Army estimators usually charged that the Soviet naval air forces were counted twice in both the air and naval figures). But the US estimates of Soviet ground forces—the bulwark of the Soviet armed forces and the pet of the conservative marshals, nearly all of whom had been ground force commanders in World War II—remained unchanged at 2,500,000 throughout the period. No increase whatever was reflected in current estimates of the ground forces from 1948 to 1955—a fact which disproves, at least for once, the contention that intelligence produced by the separate military services is always self-serving.

Why, when there were literally years to accumulate the supporting evidence, did the estimators fail to raise their estimates of Soviet ground force strength? Those who have attempted to reconstruct what went wrong have never quite been able to say. Insofar as it can be reconstructed from memory (little is preserved to explain the rationale of such estimates), the following factors appear to have been the causes of the excessive conservatism:

- It was judged, when it was apparent in 1950–51 that the strength of the Soviet forces was increasing, that earlier estimates had been too high and that therefore the increase should be "absorbed."
- Order of battle data (except for Soviet units in Eastern Europe) was too tenuous to establish with any certainty the existing number of ground force divisions, let alone what the strength of given units might be (from cadre to full TO&E).
- Because of this inadequacy of OB data, there was no demonstrable place in the Soviet ground forces to put the increase—or, in OB terms, there was nowhere an increase could be "accepted."
- The estimators themselves were essentially cautious and conservative and probably, perhaps almost unconsciously, deemed it better to let well enough along than to "rock the boat" and attempt to explain the problem to their superiors.
- And, finally, there was probably no pressing demand from the command or policy level to come up with new estimates so long as there was no material increase in Soviet ground forces in the forward areas where they might pose an immediate threat to US or NATO forces.

In retrospect, after Khrushchev's speech, an Army Intelligence "reanalysis" came up with a figure for the ground forces from 1951 to 1953 of 3,400,000—an increase of 900,000 which was quite likely still too conservative, and still short of Khrushchev's claim. On the other hand, a separate analysis of Soviet armed forces strength done independently by another agency in 1969 concluded that total Soviet military strength, minus the security forces, had totaled over 6,300,000 in 1952—of which about 5,000,000 were either in ground force line units or in other elements of the armed forces (AAA units and command and support) normally counted in ground forces figures. The latter analysis—while tacitly accepting much of what Khrushchev had said—actually concluded that his figure had been too low for the peak of Soviet strength reached in 1952–53 during the Korean war.

In summary, the spread of figures for the strength of the regular armed forces of our major adversary (and major intelligence target) for 1952–53 was:

Current accepted figure at the time	3,900,000
Army reanalysis (1960)	4,825,000
Khrushchev's figure (for 1955)	5,763,000
Separate analysis (1969)	6,305,000

Lest the reader be misled into assuming that such problems are unique to the USSR, let us examine a more recent and actually more critical example from Vietnam.

In 1964—even prior to the Tonkin Gulf incident of 2 August 1964 and well before the start of US bombing of North Vietnam in early February 1965 and the introduction of the first US combat troops in March–April 1965—North Vietnam had begun the augmentation of its armed forces and the preparations for the introduction of regular North Vietnamese military units into South Vietnam, as opposed to small groups of infiltrators which had previously been dispatched to the South. In the spring of 1964, the training of the first units (regiments) for dispatch to South Vietnam was started and the process of upgrading NVA brigades to divisions was begun. By the fall of 1964, a recall of reservists—particularly NCOs—to the armed forces was in progress, primarily to provide the trained cadre for new units and an expansion of the armed forces. But all this was under way in virtual secrecy from us, in an area where our military information had long been sparse, collection slow and difficult, and our order-of-battle data delayed and inadequate. In early 1965, however, information began to come in—much of it in the form of announcements by the Hanoi press and radio and not until later from prisoners in South Vietnam—which suggested that a substantial augmentation of the North Vietnamese armed forces was under way.

Between January and April 1965, Hanoi: issued a decree calling for all-out efforts to build up powerful armed forces, including the regular army, and to "get ready to fight"; referred in the press to "tens of thousands of newly enlisted army men" who had volunteered to prolong their service terms; announced new mobilization measures which were not fully explained but which included revision of the military service law "to strengthen building of the armed forces" and the extension of terms of service; halted virtually all releases from the armed forces; announced a campaign under which youths by the thousands were said to have volunteered to join the army and go anywhere needed to fight; announced the adoption of other decisions concerning mobilization of the country, details unspecified but intended to cope with the "new situation" in which the whole country was in a state of direct hostilities with the US and to "strengthen the building of armed forces . . . to cope with all war expansion schemes of the US."

In the ensuing late spring and early summer months, there were accumulating indications that a serious mobilization effort was in progress in North Vietnam. Evidence from classified sources augmented the material from the North Vietnamese press exhorting the people to greater efforts and calling for the youth to volunteer for the army. On 20 July, a US press dispatch from Saigon quoted Professor P. J. Honey, a British authority on Vietnam, as stating that there was a "massive military callup in North Vietnam." Basing his analysis almost solely on statements in the Hanoi press (since he had access to little if any classified material), Professor Honey concluded that North Vietnam was preparing to commit a large number of troops very suddenly in the South Vietnamese highlands. While official figures on North Vietnamese military strength reflected almost no increase, an indications roundup of the classified and unclassified evidence on North Vietnamese mobilization concluded in early August that: "The question is not 'Is North Vietnam augmenting its armed forces?' but is it doing so on a scale so as to permit an early and major introduction of additional forces into South Vietnam?"

Throughout the following months the argument continued, with indications analysts piling up scraps of information from the North Vietnamese press, prisoner interrogations and other sources to support their contention that a substantial augmentation of the North Vietnamese Army was under way, the objective of which was to put thousands of combat infantry troops into South Vietnam—the battleground on which North Vietnam expected to win the war. Meanwhile, by early 1966, the processes of order of battle analysis had "accepted" the presence of several North Vietnamese regiments in South Vietnam, thus permitting some increase in estimated NVA strength, and OB had "confirmed" the upgrading of *one* NVA brigade to division. (It had actually been upgraded in 1964, and by early 1966 there was some evidence from prisoners and/or captured documents that three other NVA brigades had been upgraded to divisions. So great was the time lag in obtaining OB "confirmation" that some units actually were being downgraded by OB to brigades in 1965, based on information up to two years old, when in fact they had already been reconverted to divisions. And it would be some time before we would understand the process of regeneration of North Vietnamese units in the north as regiments were dispatched south, the formation of training divisions, and other measures for the continuing recruitment, training and infiltration of units to the South.)

The discrepancy between the order of battle and indications approach to this mobilization will be clearer from the following: In March 1966, the official estimate of North Vietnamese Army strength was 274,000, an increase of 31,000 since July 1965, which in turn was a slight increase from early 1965. Independent, and admittedly not very scientific, indications analyses had concluded, on the other hand, that an increase of some 200,000 to 250,000 men in the NVA during the past year was not unreasonable and might even be conservative; total NVA strength based on information other than rigid order of battle criteria, was assessed at about 450,000. Interestingly enough, little more than a year later, a figure of about 450,000 was officially accepted, based on evidence which was then more than a year old. And a study done years later, based on much more evidence on the scale and timing of North Vietnamese conscription, came to almost the identical conclusions as the indications study of March 1966.

These histories of two major mobilization problems of recent years from widely separated areas have been explained in some detail so that the reader can better perceive the difference between the conventional military approach and the indications analysis. If the requirement for order of battle "confirmation" is too slow and hence may be dangerously misleading when units are deploying, still less can conventional order of battle methodology be expected to cope with mobilization. "Proof" of the formation of new units, particularly in a rapidly developing situation when deployments may also be under way, is extremely difficult to come by—it may not be just weeks, but often months or even years before the order of battle developments in a major buildup can be reconstructed. It may be just as difficult and time-consuming to determine that an under strength unit is being upgraded or that auxiliary support units are being mobilized or command echelons beefed up. When mobilization is in progress, we are almost certain to be running behind the facts—except in those cases where we are privy to the mobilization plan and we know for sure that it is being implemented, circumstances which rarely if ever obtain in closed societies.

In both the examples cited above, the situation actually could have been much more serious and we are fortunate that the gross misestimates did not cause more problems. The Soviet Union quite probably never intended to commit its ground forces during the Korean war, but we cannot be positive of this had the war gone worse for the Chinese Communist forces in Korea. In the case of North Vietnam's mobilization, it was the slow pace of the war itself, the time lag between recruitment of men and the piecemeal commitment of units to the South, the long overland trek and the extended logistic routes which permitted us to learn gradually how much effort and manpower North Vietnam was expending so that the full impact of the mobilization effort was never suddenly brought to bear against US forces—as it might have been in a conventional military situation.

The Concealment and Detection of Mobilization

In neither of the preceding examples did the nation in question make any particular effort to conceal its mobilization. The North Vietnamese in fact were announcing theirs, primarily to their own populace and only incidentally for our benefit. The Soviet effort, although not publicized as such in the

USSR, was supported by a substantial amount of information, in particular the two-class callup.

In other instances of recent years as well, Communist nations have announced, or all but announced, partial mobilization measures. Communist China's entry into the Korean war was preceded by public calls for "volunteers." (Whole units, rather than individual "volunteers" of course were actually sent, but the announcements probably also were intended to encourage enlistments.) At the height of the Berlin crisis in the summer of 1961, the USSR announced that it was holding in service "necessary numbers" of men due for release from the armed forces, allegedly until a German peace treaty was concluded. This announcement undoubtedly was intended in large measure for political impact and was part of the USSR's psychological warfare campaign over Berlin. Nonetheless, the retention of the men was a real preparedness measure, and we know from classified documents that the USSR also instituted a partial callup of reserves and a reinforcement of units in the military districts bordering Eastern Europe.

In the Czechoslovak crisis in the summer of 1968, the Soviet Union announced the callup of reservists and requisitioning of transport equipment under the guise that it was conducting a Rear Services "exercise." To lend verisimilitude to this claim, the Soviet press carried daily comments on the progress of the "exercise," but the concurrent major buildup of forces along the borders of Czechoslovakia made it doubtful that any exercises were in progress and supported the likelihood of a bona fide partial mobilization, although admittedly of undetermined magnitude.

These instances raise some interesting questions—particularly for warning—about how far we may reasonably expect any nation to attempt to conceal its mobilization efforts and what types of measures it may take to attempt to disguise or explain away mobilization. And our answers, or at least tentative answers, to these questions will help us to understand what we may reasonably expect to know—and not to know—when a nation is mobilizing.

It would seem doubtful that any nation, no matter how rigid its control of press and populace, could or would even try to carry out a full mobilization for war in total secrecy. Such a move would of course be impossible in a democracy, but dictatorships also need the support of the populace for war and some explanations of massive and sudden recalls of men to the armed forces would probably be forthcoming. Such massive military preparations also would be discernible both to natives and to foreigners, and rumors probably would spread like wildfire through the secret grapevines which exist to provide the "truth" to the populace of countries denied the benefits of a free press. We should thus expect to learn something, and perhaps a great deal, but how much we could never be sure in advance. A very rapid mobilization in preparation for imminent military action, particularly if covered by announcements of some type of major tests or exercises, could be quite deceptive, and we might well find out too late how extensive and real the effort was.

Partial mobilization measures, particularly when carried out in the border areas of large nations some distance from their capital cities, can very well be concealed. We should be under no illusions as to our limited capabilities to detect even quite extensive military preparedness measures, including mobilization, in areas denied to our observation. The capabilities of great Communist nations, and some of the smaller ones as well, have been amply demonstrated in this respect. Had the USSR not chosen to announce its partial mobilization in the summer of 1961, we would have known little if anything of it. Similarly, the announcement in July 1968 of the callups for the impending Rear Services "exercise" rather than any independent evidence was our tipoff that a partial mobilization was in progress. Although we are not sure why the USSR chose to make the announcement, a likely reason is that it knew it could not conceal the forward troop movement into Eastern Europe and wished to announce an "exercise" in advance as cover for this. There was virtually no leak through Soviet sources of the progress or extent of this mobilization at the time, and it was only in retrospect that its full extent was appreciated.

There is no doubt that the USSR, Communist China and North Korea—to name the most conspicuous candidates—are fully capable of carrying out partial mobilizations in almost total secrecy, if they choose to do so. Such evidence as we might receive, even under favorable circumstances, could well be mistaken for exercises. Troop deployments, particularly major ones, might be detected, but initial mobilization measures quite likely would not. Our capabilities in the Eastern European countries—by virtue of their smaller size and our more adequate collection and greater freedom of travel—are considerably better, but there is always the danger that drastic security measures would greatly impair our capabilities at the crucial time.

This gloomy view is reinforced by the fact that we know that most Communist states envisage the possibility of a partial mobilization in secret, and perhaps even the initiation of total mobilization without publicity. They have recognized the value of such secret preparations both for the achievement of strategic surprise and for the tactical advantages which accrue in the initial phase of the war to the nation which has accomplished in advance the mobilization of the units to be committed in the first operations of the conflict. Since doctrine calls for this, or at least for this contingency, we may be relatively sure that careful plans exist for the implementation of such secret mobilization measures and that security and deception measures have been or would be devised to mislead us.

It is because such capabilities and plans exist that it is all the more important that intelligence analysts recognize and be alert for the clues which might lead us to understand that mobilization had begun. The folly of expecting conventional order of battle methodology to provide an answer to the extent of mobilization, or even to be able to "confirm" that mobilization has begun at all, should be apparent. We must be prepared to accept more imaginative, if less exact, methods.

Our ability to recognize that mobilization has begun, if the enemy wishes to conceal it, will likely be dependent in large measure on related developments. Most important could simply be the existence of an international crisis which would make partial mobilization a likely move. If this situation is correctly perceived, various military anomalies—such as cessation of normal training, indications of alerts, sudden shortages of transport for normal civilian purposes or sudden appearances of additional transport and combat equipment in installations of military units may provide clues that a genuine mobilization is under way. The recognition of the possibility will likely also call for increased or extraordinary collection measures which may provide more evidence.

In addition, it will be extremely important to understand correctly what the enemy is saying. The fact that both North Vietnam in 1965 and the USSR in July 1968 virtually announced their mobilization in transparently obvious statements did not mean that all or even a majority of analysts perceived it. The community, and especially military analysts not particularly experienced in the analysis of propaganda, need to acquire a better understanding of what such statements may mean. Many seem to go to one extreme or another—they accept all statements at face value or they reject them totally as "propaganda." More careful study and analysis of enemy statements is needed, and the assistance of experts on these matters should be sought. Too often the enemy is really telling us what he plans to do or is doing and we are reluctant to perceive or to accept it.

Students of warning as well as basic military analysts should take time to read and study the basic legislation, which is often unclassified, which governs the terms of service for men in the armed forces and often the conditions under which they can be recalled or held in service beyond their terms. It is a fundamental error to consider that such legislation is so much window dressing or that Communist nations never observe their laws. In fact, these laws are applicable and nations such as the USSR often take particular pains to observe, at least on the surface, the terms of their legislation. The public

Soviet decree that men were being retained in service in 1961 may have been predicated in large part on the fact that the law prohibits such extension beyond two months without appropriate action of the Supreme Soviet. It was reported after the invasion of Czechoslovakia that some Soviet reservists went on a hunger strike in early August when their three months' legal recall to active-duty had expired (these were the first group of reservists called up for the initial deployments of Soviet troops in early May). As a result, they were reportedly released and a whole new group of reservists called up only four days before the invasion, at least in one unit. An understanding of basic but rarely needed facts of law, doctrine, policy and practice may well assist the analyst in time of crisis to perceive that a mobilization has begun.

For the most important thing for warning is to recognize that a mobilization—full or partial—is in progress. If we are to be alert to the possibility of impending hostile action and so warn our superiors—we must be willing to accept the fact of such mobilization even when we cannot demonstrate its extent or be able to "prove" that a given unit or a number of units have in fact been mobilized. To recognize the fact of mobilization, we must often rely on indirect evidence—including public statements of the enemy which are often most revealing, and deductions based on partial information that some reservists or groups of reservists are being mobilized. In some cases, we may never be able to confirm the extent of the mobilization, or there may be widely varying estimates even though some compromise may finally be reached on an "official" figure. But this is less important, far less important, than that the appropriate authorities be warned that some mobilization, extent not yet clear, is in fact under way.

Notes

1. Barbara Tuchman, *The Guns of August* (New York, MacMillan Company, 1962), pp. 74–75.

2. In an historic statement, which passed virtually unnoticed, a Soviet broadcast to China on 21 August 1970 finally acknowledged the USSR's participation in the Korean war "The Soviet Union sent military advisers to Korea. They directly took part in combat. . . . The Soviet Union dispatched air force units to China. Soviet pilots downed scores of US planes and protected northeast China."

16

Logistics Is the Queen of Battles

IT WOULD BE IMPOSSIBLE in a single chapter to give adequate attention to the importance of logistics for the conduct of war—and hence the importance of logistic indications and the detection of logistic preparations for warning. The measure of the significance of logistics is conveyed in the title of this chapter. Simply stated, logistic preparations are of decisive importance, and the level of logistic activity can be the single most important indication of the likelihood, or unlikelihood, of hostilities. It may be, and often is, *the* determining indication that a real military buildup for combat is in progress, as opposed to exercises or a show of force. The logistic buildup *always* will be greater for combat than for exercises.

The extent and variety of logistic preparations for modem war are reflected in the number of logistic and transportation items carried on indicator lists, which usually equal or exceed the number for any other topic. An indications specialist, asked where he would most wish to have an espionage agent in the military establishment of a potential enemy state, replied, "In the office of the Chief of Rear Services"—the name by which the military logistic and support services are known throughout the Communist world. For if we could be sure of knowing the extent, level and variety of logistic preparations at any time we would not only have a very accurate grasp of the enemy's capabilities, we would probably also have very precise insight into his intent.

In this chapter we will be considering only those aspects of support, supply and movement which are an integral part of the military establishment or which are indirect support of military operations—as opposed to more general and largely civilian economic indicators (which are discussed in Chapter 22).

Nations Are Not Logistically Ready for War

So much has been said in recent years about the threat to us of our potential enemies, the extent to which they allocate their resources to military preparations, and how many missiles, tanks and aircraft they are producing that it is easy to gain a mistaken impression that these nations also are in a high state of readiness logistically for hostilities, and would need to undertake little additional preparation for war. At one time, at least, and even to some extent today, such views have been rather widely held, particularly with regard to the Soviet Union.

As with many other things, some experience with live crises and improved collection have given us a better appreciation of reality. It is true, of course, that the USSR does produce large quantities of weapons and equipment and it does have substantial stocks of many items in reserve depots—some in forward areas and others farther to the rear—which would be drawn upon in the event of hostilities. In the forward area (i.e. Eastern Europe) there have been rather widely varying estimates of the quantities of supplies on hand; some have been judged adequate for combat for a month or more, others for less, and some items appear to be in quite short supply, including some which might be most crucially needed in the event of actual hostilities.

In any case, whatever our estimates are, the USSR itself has clearly felt its logistic preparations to be inadequate for the possibility of hostilities. Not only did the Czechoslovak crisis reveal a requirement to mobilize Soviet combat units, still more striking was the requirement to mobilize Rear Services support units. In order to support the forward movement of combat units, it was necessary to requisition transport vehicles and their drivers from their normal civilian activities—and this at the height of the harvest season when they were most needed. These reserve transport units were then employed in shuttling supplies from the USSR to forward bases both prior to and after the invasion of Czechoslovakia and were only demobilized in the autumn when the situation had stabilized sufficiently to permit a withdrawal from Czechoslovakia of a substantial portion of the original invasion force. There are some indications that, even with this effort, supply shortages were encountered—although there was no active resistance.

In addition, the USSR took some extraordinary supply measures in Eastern Europe itself. For the first time within memory, there were reliable eyewitness observations of the emptying of ammunition dumps—one divisional ammunition facility in East Germany was stripped even of the guards on the watch towers, an event believed to be unprecedented since World War II. These incidents had occurred before 1 August, or three weeks prior to the invasion.

These points are made here to emphasize that these activities bore no similarity to an exercise (although in a pro forma deception effort the Soviets called much of their logistic activity an exercise) and were clearly distinguishable from the type of activity normally conducted in Eastern Europe. To those analysts who were convinced of the likelihood of a Soviet invasion, the logistic preparation above all was perhaps the most decisive evidence. Even more persuasive than the deployment of twenty plus divisions of Soviet troops to positions around Czechoslovakia (which conceivably, albeit with some difficulty, could have been explained away as mere "pressure" on Czechoslovakia) was the reality of the logistic buildup. There was no conceivable need for it except for an actual invasion.

This is but one of innumerable examples which could be found to demonstrate the validity of logistic preparations as a barometer of the enemy's preparations for hostilities. Numerous instances could be cited from the Vietnam War—ranging from such relatively long-term preparations as the construction of new roads in the Lao Panhandle to handle the truck movements to the South, to such short-term tactical preparations as the commandeering of the local populace to porter supplies in preparation for an attack on a fortified village. In large part the history of the war has been a chronicle of ingenious and unrelenting North Vietnamese efforts to sustain their logistic movements and of our attempts to disrupt them. With few exceptions, major new logistic projects or exceptionally heavy movements of supplies have proved to be valid and reliable indications of enemy preparations for forthcoming operations.

In short, the type of logistic preparations undertaken by any nation in expectation of early hostilities is different in both quantity and quality from what goes on in time of peace. And we can—if we have enough evidence and understand our adversary's methods of operation—usually see the difference.

Key Warning Factors in Logistic Preparations for Combat Operations

From the multitude of potential logistic indications of impending hostilities—some of which may be quite specific for particular nations—we may generalize on several of the more important aspects of logistic preparations, some or many of which will nearly always be undertaken prior to military action. Clearly, the extent and variety of such preparations will be dependent on the type and scope of expected hostilities, their likely duration, and the degree of counteraction which is anticipated. A nation undertaking a relatively small operation involving little risk against an ill-prepared adversary obviously will need to undertake fewer and less extensive preparations than a nation preparing large numbers of forces for major operations against a formidable opponent. The scale of the logistic preparation alone, provided it can be determined, thus may provide very good clues as to how extensive the impending operation is likely to be.

a. Logistic preparations an integral part of mobilization plan. When the impending military operation requires any degree of mobilization—as it usually will—the mobilization also will involve some type of logistic preparations, since the two are inseparably connected. Reserves called up must have weapons and ammunition, expanded units must obtain or remove equipment from depots, additional transport (rail cars, trucks, aircraft and sometimes ships) is needed to move both troops and their supplies and equipment, more POL is needed both to move the units and to support them in the impending action, and so forth. Even the mobilization of a single reserve or under strength unit in a modern army will require a whole series of logistic support measures if the unit is to move anywhere or have any combat capability when it gets there. The more troops involved, the more extensive the logistic and transport support which will be needed—and the more disruptive or apparent the fulfilling of this requirement is apt to be. Logistic planning for the mobilization and deployment of modern forces is enormously complex and, indeed, the greater part of the mobilization plan is actually composed of the details by which the supply, support and transport of forces is to be accomplished rather than the mere callup of the reservists themselves to their units. The more carefully the mobilization plan is prepared and the more rapidly it is to be implemented, the greater will be the meticulous attention to every detail of logistic support, since any shortage or bottleneck can disrupt the whole system.

b. Impact on civilian life and economy. The mobilization of additional logistic support on any scale is likely to have an early and sometimes serious impact on the life of the ordinary civilian. Certain factors of course will mitigate these effects: well-placed and large reserves of military supplies and equipment may permit a unit to mobilize locally with no seeming impact on the community; some types of foods can be preserved for long periods; POL depots reserved exclusively for a military emergency will reduce the need to requisition it from the civilian economy; and very affluent nations (such as the US) may have such a general abundance that a relatively small mobilization will have little or no immediate impact on civilians.

Nonetheless, all nations have some bottlenecks and shortages, and in a mobilization of any scope the effect of these will be felt, and often immediately, on the life of civilians. Warning analysts should be cognizant of what these potentially critical shortages are and collectors should be prepared in advance to concentrate on some of these—since a coincidence of several of them may be either long-term or short-term evidence of mobilization or of unusual military requirements. Some of the types of shortages indicative of military requisitioning which have shown up most frequently in the Communist world have been: *food* (particularly meats and other choice items which are nearly always diverted to the military in time of need); *POL* (despite reserve depots for the military, crises involving military deployments are usually accompanied by reports of POL shortages for civilian use); and *transportation* and *medical supplies* both of such critical importance that they are further discussed below).

c. **Impact on transportation.** One of the most immediate and most disruptive effects of logistic mobilization measures is likely to be on the transportation system—and happily for us it is also one of the most likely to be detectable. No nation has a transportation system adequate to take care of any substantial increase in military requirements without either requisitioning from or otherwise having some effect on normal civilian transport. The mobilization plans of all nations call for the commandeering in one way or another of civilian transport facilities. These measures may range all the way from simply giving priority to military shipments (by rail, truck, water or air) over civilian shipments up to extensive requisitioning of trucks from the civilian economy and/or the takeover by the military of the entire transportation system.

The most extensive mobilization plans in the world for the takeover of civilian transport by the military are probably those of the USSR and its Eastern European allies (which have closely integrated plans). Some of these nations also suffer from acute transportation problems and difficulties, including shortages of rail cars and trucks. The result is that any significant or sudden increase in military requirements will usually have an immediate impact on the transportation system. Reports of shortages of rail cars for normal use have sometimes been a key indicator that abnormal military movements were impending—e.g., for the buildup of forces around Czechoslovakia in the last week of July 1968. Moreover, we have learned that many (probably most) ground force units lack their organic truck transport in peacetime and must obtain it from the civilian economy when they mobilize (this was also done for the invasion of Czechoslovakia). The mobilization plans of most Soviet and Eastern European units for a long time to come will probably require a requisitioning of trucks from elements of the civilian economy which also have a critical need for it—particularly agricultural enterprises.

For a major mobilization in expectation of large-scale hostilities, these nations plan a virtual total takeover of the transportation systems—ground, air and water. Since these systems already are wholly state-owned and controlled, they are fully integrated into the mobilization plan. Key transportation officials, as well as many civil airline pilots, are reserve officers. All Soviet merchant ships have reserve officers aboard, and have rehearsed their mobilization plans. The truck transport units which will be requisitioned from the economy have been identified in recent years as reserve military units; their equipment is better than average and they periodically practice their mobilization plans. Extensive plans undoubtedly exist to place Soviet military railroad officials in the rail systems of Eastern Europe in event of mobilization, if not for the Soviet Army to take over the system. In short, the entire wartime transportation system is geared to go into effect on very short notice, although how effective all this would actually be in a sudden crisis is somewhat less certain.

However effective or ineffective the implementation of these plans, we could hardly fail to detect some aspects of it, particularly in a drastic and sudden military takeover accompanied by heavy military use of rail and road transport into Eastern Europe. A Soviet military writer some years ago was so concerned over the detectability of sudden increases in military movements into Eastern Europe that he proposed that reinforcements be moved in gradually and the normal pattern of freight movements continued as far as possible. This is certainly a possibility which should be watched and could well be part of a long-term secret plan for surprise attack, if in fact the USSR should ever come to consider such strategy as desirable (and non-suicidal).

In other key areas as well, major disruptions of transportation as a result of military movements have been key indications of preparations for hostilities. The heavy Chinese Communist troop movements to Manchuria in the summer and fall of 1950 caused extensive disruption of normal freight movements—and there were a number of reports of this. Less extensive troop and supply movements in China also have occasioned sporadic reports that civil transport has been curtailed, and this seems to be one of our few relatively reliable and potentially detectable indications of troop movement in mainland China.

d. Heavy movements of rear service elements and supplies. However many supplies may have been stocked in forward depots or in reserve to the rear, no commander probably ever thought he had enough. The additional buildup of supplies for the forces to be committed in initial operations (whether they be ground assault forces, air, naval, air defense or even missile forces) is a virtual certainty for any nation which has time to make such preparations. Unless forces are taken by total surprise, some logistic buildup prior to operations is to be expected—and usually it will be very heavy. In fact, the logistic buildup often will be as heavy and demanding of resources as the troop buildup, if not more so—and, moreover, will require a longer time. The problem of the "logistic tail" is well known, and many a commander has been ready to go before his supply buildup has been completed, or has been forced to halt his offensive for lack of such key items as POL.

A delay in attack after the combat forces are seemingly ready and in place therefore may not mean any hesitancy or indecision on the part of the enemy so long as the supply buildup is continuing. It should not be regarded as a negative indication, as it so often has been. Rather, a most close watch should be maintained on the logistic buildup in the hope of determining when in fact the enemy will be ready to jump off. (This may not be easy to determine, but it will probably be somewhat better than total guesswork.)

In World War II, the Japanese watched the buildup of Soviet forces in the Far East for about six months. They had estimated that the facilities of the Trans-Siberian Railroad would have to be concentrated on military shipments for at least 100 days before a Soviet attack on the Japanese forces in Manchuria. By late July 1945, their intelligence was reporting that continued heavy movements now consisted almost entirely of supply and rear service troops, which they correctly interpreted to mean that the buildup of combat forces was complete. They concluded that the Soviets could initiate the attack at any time after August and received no further warning of the attack which occurred on 8 August.

In the Soviet buildup for the attack on Czechoslovakia in the summer of 1968, most of the combat forces which participated in the invasion appeared to have completed their deployments by 1 August (this excluded the backup forces which entered Poland from the Baltic Military District whose movement was still in progress well into August). The intelligence judgment was that Soviet forces were in a high state of readiness (if not their peak readiness) for invasion on about 1 August. In fact, however, the logistic buildup in Europe was still continuing and the Soviets themselves did not announce the completion of their professed Rear Services "exercise" until 10 August. It is likely that Soviet forces really were not ready, or at least not as ready as their commanders would have wished, at the time that the community judged them to be. Had there been a real threat of serious opposition, the logistic buildup probably would have been of even greater importance and presumably might have required a longer time. In any case, the relaxation of concern over a possible invasion of Czechoslovakia when nothing occurred in the first few days of August was not warranted by the military evidence. The threat not only remained but was increasing.

The problem of assessing logistic readiness, and hence actual readiness for attack, has been a difficult intelligence problem in Vietnam and Laos, where the problems of estimating supply movements at all have been compounded by the effects of air strikes, road interdictions, poor weather and jungle terrain.

In these circumstances, however, as in more conventional warfare, the continuing effort to buildup supplies and the evident great importance attached to keeping the roads open have been indications of impending operations even when their scope or timing might be elusive.

e. Drawdown on existing depots. Another aspect of the abnormal requirement of military units for supplies is likely to be the drawdown on supplies in depots, both in forward and rear areas. In the forward area, it will be for allocation to the troops for combat (as we have noted, the unprecedented emptying of ammunition dumps in Eastern Europe prior to the Soviet invasion of Czechoslovakia was a strong indication that the basic load of ammunition either had been or would be issued to the troops). In rear areas, a sudden emptying of reserve depots, particularly if several are involved, is a likely indication that equipment is being issued to reserve or understrength units.

f. Medical preparations. One of the most obvious requirements of forces going into combat as compared with their needs in peacetime is in medical support. Indeed, of all fields of logistic support, this may be the one in which there is the greatest difference between war and peace. Not only will combat units require vastly greater quantities of drugs and medicines, but also different types of drugs. The requirements for doctors and medical technicians will multiply, as will the need for field hospitals, evacuation facilities, etc. Even in nations which do not traditionally place a high value on saving the lives of wounded enlisted men (among which have been several important Communist nations), the need for increased drugs and medical support facilities of all types will skyrocket in the event of hostilities.

As in other logistic preparations, these changes will affect not only the military forces themselves, but also are likely to have a serious impact on the civilian populace. Shortage of drugs and hospital supplies, sudden callups of physicians to the armed forces, and greatly stepped-up blood collection programs are some of the obvious effects.

In addition, belligerents may have to make very heavy purchases of drugs abroad (as did North Vietnam, assisted by Communist China, in the early stages of US involvement in the Vietnam War). Occasionally, heavy acquisition of drugs for particular ailments (e.g., anti-malarial) may provide a tipoff that military forces are being readied for movement into particular areas.

Some warning analysts feel that medical preparations are an aspect of war planning which have been given too little attention, and that this is a field in which our data base and collection planning could well be improved against the day when one of our major enemies might make drastic changes in its medical practices in preparation for major hostilities. It is a field in which some preparedness measures would appear to be unique to hostilities and thus of very high specific value for warning.

g. Logistic preparations for survivability. Another important logistic preparation likely to distinguish war from peace is the stockpiling of equipment and other measures to ensure the survivability of the logistic system itself. Such measures, particularly if undertaken in any great number or in marked contrast to normal practice, carry high validity as indications that the enemy is preparing for some action for which he expects retaliation. (Unless, on the other hand, we ourselves or some other nation are clearly getting ready to initiate the action.) Rail and road facilities are, of course, highly vulnerable to interdiction, and aggressors normally will take some measures to protect them and to provide for their repair before they embark on hostilities. In addition to obvious measures such as increased AAA protection at bridges and rail yards, extraordinary logistic preparedness measures will be suggested by the stockpiling along the way of railroad ties and rails, emergency bridging equipment, bulldozers, road-building machinery and other material for the rapid repair of damaged transportation routes. Rarely, if ever, do nations undertake emergency steps of this nature when they are expecting a prolonged period of peace. Another measure to ensure the survivability of critical transportation routes is the construction of numerous road bypasses, extra bridges, etc. at bottlenecks—a procedure at which the North Vietnamese have been adept.

h. Logistic preparations for specific operation. In addition to the value of logistic preparations as indications in general, certain types of activity may provide very specific indications of the nature of impending attacks. An extraordinary standdown of transport aircraft or their assembly at bases near airborne forces is a good indication that airborne operations are impending. A large buildup of amphibious equipment and landing craft probably signifies impending water-borne attacks, and the detection of large quantities of mobile bridging equipment probably indicates major river-crossing operations. These highly specific logistic operations are of great value for warning, provided they are detected, since they give us not only generalized warning but often specific warning of how and where attacks may be launched. Thus, once again, we can demonstrate the value for warning of the identification of the enemy's logistic preparations.

Interpreting Longer Term Logistic Preparations

In the foregoing discussion, we have dealt with relatively short-term logistic preparations—those associated with a buildup of combat forces for a specific operation. The intelligence analyst will also find, however, that he must come to some interpretation concerning a number of longer term logistic activities whose significance and relevance to warning are less clear-cut.

All nations are engaged in improving their transportation facilities, and most military forces also seek to build up their basic facilities (sometimes known as "infrastructure") and to augment and modernize their stocks of equipment. Unless halted by politicians or budgetary restraints, it is generally considered that military commanders will constantly be acquiring more materiel and stockpiling more, and thus continually building up their capabilities. We frequently receive reports of new POL depots in Eastern Europe or of road and rail construction which would increase military capabilities in event of war, and the Communist nations of the Far East also are engaged in such programs.

What do these measures mean for warning? In many cases, the only answer which we can make is that these improvements do enhance enemy capabilities and also, in some cases, appear to be long-term contingency preparations. Beyond this, there is usually not much which we can say for warning purposes. Some of these developments are strictly military (depots, for instance), while others also serve a legitimate economic function and might have been undertaken in any case. We need to watch these preparations, to make sure that we have not missed something or that they do not appear greatly excessive for any immediate need, but beyond that we are not usually very concerned about them.

There are times, however, when some of these basically long-term logistic preparations, such as new road construction, may have indications significance. Obviously, North Vietnamese road construction in the Laotian Panhandle has had the most direct bearing on Hanoi's military capabilities and intentions in South Vietnam. There have been few items of greater indications significance for the US in recent years than this extensive road construction effort.

Less certain has been the meaning of the Chinese Communist road construction program in northwest Laos. Is this a long-term project designed only to improve transportation in this remote area, or is it preparatory to the support of military operations in Laos or Thailand?

Or, in another critical area, what is the import of the buildup of Soviet logistic facilities, airfields and so forth near the Chinese border, which has accompanied the buildup of combat forces under way in that area since 1965? Do these developments indicate Soviet preparations for hostilities at some time in the relatively near future, or is it all only a long-term program to be prepared for any contingency?

There are no clear-cut guidelines as to when we can conclude that preparations of this nature are for support of early hostilities and when they are not. Our conclusions must be tentative and subject to change as more information becomes available. There are certain general guidelines which may assist the analyst, however, some of which are:

Urgency. If the construction effort is relatively slow, and particularly has extended over a period of years when it might have been accomplished more rapidly, it is hard to believe that it is related to some specific plan for military operations which the nation plans to implement when the construction is finished. On the other hand, if the effort does appear urgent and particularly if it has diverted resources from other important projects, there is a good possibility that it is for some relatively short-term purpose.

Other uses. If it is urgent and there is no valid economic or other civil purpose for the project, then the chances will be further increased that its purpose is military, and potentially ominous.

Is it excessive? This will be a most important factor in assessing its military importance. Marked buildups of logistic support facilities greatly in excess of seeming requirements of forces already in the area will tend to raise a flag; they suggest that the stage is being prepared for the arrival of more forces—and perhaps very rapidly after the logistic support buildup has been completed. On the other hand, a logistic support program which about keeps pace with the normal requirements of forces in the area—even though the latter are expanding—must be regarded as "normal"—i.e., we would expect large standing forces to have such logistic support.

How permanent is it? A military buildup which is accompanied by large-scale permanent-type construction (standard barracks, depots, major airfields, facilities for recreation or dependents, etc.) is usually a long-term repositioning of forces rather than a preparation for early hostilities. One does not construct permanent barracks and depots for troops which are only staging through to attack someone. These troops and their supplies will be housed in tents or other temporary facilities—or put up with civilians.

What else is going on? In the end, our assessment is likely to be dependent above all on our overall judgment of the likelihood of hostile action in the area. If the general atmosphere is peaceful and the nation in question seems clearly preoccupied with domestic matters and seemingly has no immediate desires to grab any territory, we almost certainly will also judge the logistic preparation to be of no immediate warning significance. And the chances are high that we will be right. Few would be likely to conclude only from a few logistic preparations, no matter how seemingly useful for war, that war was imminent.

As this chapter should have made clear, there will be few if any topics so important to the warning analyst—and the accurate warning judgment—as a correct appreciation of logistic developments. One final note of guidance therefore will be offered.

Don't forget or overlook the logistic items in the mass of other indications. In a major buildup, a great deal will be going on—large-scale ground force movements, lots of fighter aircraft redeployments, dozens of political items and rumors, accurate and planted reports of enemy intentions. All these are heady stuff and make fine current items, and the order of battle buildup looks great on the map. But, sometimes in all this, those little logistic fragments get lost which might have confirmed that the military buildup is genuine—that is, involves bona fide preparations for combat and is not an exercise or an idle threat for our benefit. In all the excitement, don't lose these gems of intelligence so rare and invaluable to the assessment of the enemy's real intentions.

17

Other Factors in Combat Preparations

THE THREE PRECEDING CHAPTERS have addressed the major types of activity—troop movements, mobilization, and logistics—which precede or accompany the commitment of forces to combat. The student of warning who understands how his adversary goes about these preparations, and who is able to perceive what is abnormal or highly unusual from what is normal or routine, will be a long way on the road to recognizing the true buildup for combat.

Preparations for hostilities, however, will likely be marked by a number of other anomalies or highly unusual activities which will further help to distinguish the situation from that which is normal. Indicator lists are concerned with dozens of such items, and the reader is referred to these for more specific identification of developments which have been judged to be of particular significance for warning. In this chapter and the next, we will consider in general terms some of the things which we should watch for, or which have been noted in past crises, which characterize true preparations for hostilities.

Preoccupation of the High Command

War is the business of generals and admirals and, when preparations are in progress for war, they will be very deeply involved indeed in the planning—even to the consideration of the most minute details of the impending operation. Their staffs will be putting in long hours of planning and paper work, and more officers will likely be needed to cope with it. For reasons of security and secrecy, as few officers as possible may be informed of the full plans, but it is inevitable that the preparations will affect the activities of many, if not all, components of the military establishment in one way or another. This will be true, although in varying degrees, whether a deliberate surprise attack is being planned for some months ahead (as in Japan's preparations before Pearl Harbor) or whether the military leadership is preparing to respond to a relatively sudden and unexpected crisis (such as the US when the strategic missiles were discovered in Cuba).

One of the great myths perpetrated by some analysts (sometimes in the name of warning) is that our enemies have all their contingency plans ready and that great hostilities could start at the drop of the hat without any further planning or consideration of the details by their General Staffs. We know that we could not do that, and that every crisis (even minor ones) engenders an extraordinary level of activity and a deluge of paper work and staff conferences and enormous volumes of communications. But somehow the idea nonetheless prevails that our enemies are better organized or prepared and that they can plunge into war without all this activity, that they have a plan ready for every conceivable contingency, and that all they will have to do is push the right buttons, or send a few brief messages to subordinate commands to open up the right contingency plan, and everything will roll with no further hitch.

This is ridiculous. No nation can be ready for every contingency, and even if it does have a specific plan for the particular crisis at hand, the High Command or General Staff assuredly will want to review it and to be sure that all the subordinate commands are fully prepared to play their roles and there are no bottlenecks, etc. More likely, they will have to modify the plan or prepare a whole new one, since the situation almost inevitably will be a little different, or someone questions part of it, and so forth. In any case, the result is a virtual certainty: the military establishment at its highest levels will be extraordinarily involved and preoccupied with planning and staff work, intensive activity will be under way, the command and control system will be tested to its fullest, and the ramifications of all this are likely to be reflected in high volumes of communications and a general atmosphere of crisis in the military establishment. Provided we have some collection assets which will give us some insight into what the military leadership is doing or the amount of activity at command levels, we are almost certain to get some sense of this although it will likely be fragmentary. But in no case will the activity of the military leadership be "normal."

We have learned from experience that even major exercises are likely to require the attention and presence of top military leaders. Where the preparations are "real," the requirement for their participation will be infinitely greater. A knowledge of their whereabouts and some information on their activities may be invaluable. From a variety of sources, both military and political, we hopefully will gain some sense, however inadequate, that the military establishment is immersed in great and extraordinary activity, or is devoting an inordinate amount of attention to a particular area or subject.

Alerts and Combat Readiness

The declaration of the real (not practice) alert and the raising of forces to very high (or the highest) levels of combat readiness is a major indication of preparedness for combat or possible combat. Like mobilization or major redeployments of units, the true combat alert will be reserved only for the extraordinary situation in which hostilities are either planned or there is reason to fear that deteriorating international conditions entail a risk of hostilities in the near future.

Forces of major nations usually have prescribed degrees or conditions of readiness ranging upward from their normal day-to-day condition to the highest degree of readiness which is full readiness for imminent hostilities. In the US forces, these stages are known as DEFCONS (Defense Conditions) which range from 5 (lowest) to 1 (highest). The highest DEFCON ever declared by the JCS for US forces as a whole is DEFCON 3. Under very rare and exceptional circumstances (e.g., the Cuban missile crisis), some components of US forces have been on DEFCON 2. The raising of DEFCON status thus is a serious matter and involves specific prescribed steps which materially increase the readiness of men and materiel for combat. Imposition of the highest DEFCON status would indicate a national emergency. A somewhat modified system exists in NATO.

Our major adversaries—particularly the Soviet Union and its Warsaw Pact allies—have a similar system by which forces are brought from their normal day-to-day readiness condition to full readiness for combat. As in Western forces, these steps entail a series of specific measures by the various components of the armed forces which, when fully accomplished (i.e., when full readiness is implemented) will have brought both combat and supporting units to readiness for immediate commitment to combat. These readiness conditions are to be distinguished from routine alerts or exercises, which are frequently carried out to test the ability of units to respond quickly. As in the US forces, the raising of readiness conditions is a serious step, undertaken rarely. Indeed, there is reason to believe that the imposition of full readiness conditions throughout the armed forces, with all the steps that would presumably be entailed, is a step (like our DEFCON 1) which would be undertaken only in expectation of hostilities. Although aspects of the combat alert plan are undoubtedly tested frequently, the widespread implementation of full readiness is reserved for the genuine combat situation, or expectation of such a situation. Readiness conditions—both the terminology and the responses required—are believed very similar if not identical throughout the Warsaw Pact forces. Although probably less sophisticated, similar systems are believed to be used in other Communist states.

The declaration of full combat readiness is thus a matter of highest concern to warning intelligence. There is nothing too trivial for us to know about so important a subject: the precise terminology, the mechanics by which such a condition would be ordered, or the exact measures which follow from such an order. Because of the rarity with which even increased (as opposed to full) readiness conditions are imposed, most analysts will have had little experience with such circumstances, and there is likely to be considerable uncertainty about what we can or cannot detect or about what precisely will occur. The warning student, however, should make a study of all available evidence on this subject, for in the hour of crisis the ability to recognize that full combat readiness was being implemented could be of decisive importance.

Exercises Versus Combat Deployments

Both compilers and users of indicator lists are frequently puzzled by the fact that an abnormally high level of exercise activity and a virtual standdown of normal exercise activity both appear as indicators. This is not a real contradiction provided that the circumstances in which each of these may be valid indicators are understood.

An exceptionally high level of training, and particularly very realistic or specialized training for a specific type of operation or against a specific target, of course may be a valid indication of preparations for combat operations. Obviously, troops need training, and very intensive training, if they are to accomplish their missions effectively, and the scope and type of training in the armed forces of any nation is a very useful guide indeed to the type of operations it expects to conduct. If our collection is adequate, we may also be able to tell from specific training exercises exactly what combat operations are being planned. The North Vietnamese Army traditionally has held detailed rehearsals with sand tables before launching attacks on specific targets in South Vietnam and Laos. For larger, more conventional operations, extensive command post and staff exercises, supplemented by drills of the troops in their specific roles, are the normal procedure. No commander would choose to launch combat operations until he had satisfied himself that both his officers and troops were reasonably well trained in what they were to do, and the more specific the training for the particular type of warfare or target, the better.

Thus, as a general guide—and sometimes quite specific guide—to the plans of the potential enemy, the type of training is obviously an indication not to be ignored. Such training activity, however, is often a relatively long-term indication, which may precede the actual initiation of hostilities by weeks, months or even years. It may, in fact, not indicate an expectation of hostilities so much as a desire to be prepared for them if they should occur.

Over the shorter term, perhaps a period of some weeks although it may be much less, extraordinary changes in the pattern of training activity—and particularly cessations or near standdowns of normal training—have usually proved to be a much more specific or valid indication of hostilities. The genuine alert of forces in a crisis or for possible hostilities is nearly always marked by abrupt curtailment of routine training, usually accompanied by a recall of forces in the field to their home stations in order to place them in readiness. In a real combat alert, missile troops will not be sent to remote ranges for training nor antiaircraft units to firing areas; they will be needed either at their home stations or wartime deployment areas. Similarly, much if not all routine ground, air and naval activity will be terminated, although defensive measures such as air and naval patrols will probably be stepped up. Such variations in training patterns are likely to be evident and they are a distinguishing characteristic of genuine crises—a true sign that combat readiness of potential enemy forces is indeed being raised. The intelligence community, based on growing experience, will often be able to recognize such situations, but we need constantly to be alert to what is not happening (but ought to be happening) as well as to what is. This is particularly true when there are no overt tensions which might alert us—when the enemy is preparing in secret.

Statements that "Things look pretty quiet," or "There doesn't seem to be much going on," can be dangerously misleading. Intended as reassurance to the listener, they may in fact mean that normal activity has been curtailed because of an alert.

Once it is under way, an exercise is something of a negative indication in the short term. When troops and logistic resources have been committed to an exercise, they are not as ready to respond to a combat situation as they were before they deployed for the exercise. The problem for intelligence is to be sure that the buildup phase for an "exercise" is really for an exercise and not a cover for a deployment of forces for an attack. In many exercises, this will be little if any problem since the exercises probably will not be conducted in the area in which the troops would have to be deployed for an actual attack, but farther back from the line of confrontation or even some distance to the rear. (Soviet naval exercises designed to test the seizure of the Danish Straits are held well back in the Baltic, not up off the coast of Denmark, and so forth.) There may be circumstances, however, in which the most careful collection and meticulous analysis will be required to ensure that we can distinguish between preparations for combat and preparations for exercises.

Obviously, the most important time to be able to make such a distinction is on the eve of attack. Western intelligence now generally expects—and Soviet doctrine and practice in at least one major instance have borne out—that the USSR would deploy its forces for an attack under cover of preparations for an "exercise." The USSR and its Warsaw Pact allies similarly look with suspicion on major NATO exercises as a potential cover for aggression. The Soviets recognize that major redeployments of forces probably could not be concealed and that the best hope for misleading the adversary would be to lull him into believing that it was just another exercise. To help to lull him, the plans will probably also call for an announcement or series of announcements about the forthcoming exercise, perhaps followed by further statements on how it is "progressing."

We will be addressing the problem of this and other types of deception in a later chapter devoted exclusively to that subject. There are few deception measures which are so simple to carry out or seemingly so transparently evident as a series of false statements on the nature or purpose of military activity. Despite this, such falsehoods have sometimes proved extremely effective, and we have no assurance that the intelligence community as a whole will necessarily recognize them for what they are. During the summer of 1968, the USSR announced a whole series of exercises which Soviet and Warsaw Pact forces were said to be undertaking either in or adjacent to Czechoslovakia. In fact, these forces conducted no real exercises all summer long, and the alerting and deployments of forces for the Czechoslovak crisis totally disrupted the summer training program. These "exercises" were announced solely to provide a pretext for the introduction of Soviet forces into Czechoslovakia in June and, still more importantly, as a cover for the major mobilization and deployment of Soviet forces after mid-July. The USSR even announced an elaborate scenario for its so-called "Rear Services Exercise" under cover of which combat units and their logistic support were deployed into Poland. Yet, despite the circumstances and the obvious fact that all this activity was clearly related to Soviet efforts to bring the Czechoslovak situation under control, much current intelligence reporting accepted Soviet statements on these "exercises" as if they were valid descriptions of what was occurring. Even after the invasion, a number of post-mortem studies persisted in referring to the "exercises" without qualification, and even repeated (as if valid) the scenarios which had been carried in the Soviet press.

Our ability to tell when genuine exercises are in progress or when the training program has been disrupted or curtailed is, of course, in large part dependent on the quantity and quality of our collection. There are areas in which our coverage is so limited or so delayed that we really may not be able to tell. Our limited knowledge of Chinese Communist exercise patterns and programs would presumably make it much more difficult to distinguish an exercise from combat preparations in that area than in Eastern Europe. There would be good grounds for great uncertainty in North Korea. There could be many other areas or circumstances in which even the most perceptive and sophisticated analysis might not permit us to make a firm judgment.

But in the case of Czechoslovakia, there was ample evidence at hand to make a judgment that the Soviets were not engaged in genuine "exercises" and that Soviet tactics all summer long were designed to bring Czechoslovakia into line by one means or another. Thus the situation called for exceptional care and sophistication in reporting to ensure that the real Soviet purpose and objective was not obscured by rote-like repetition of Soviet statements. Such a perception and an accurate reporting of the real nature of things could, of course, be infinitely more important to US security in another instance.

Offensive Versus Defensive Preparations

In an impending crisis, when war may be about to break out or when there is a fear of escalation of hostilities already in progress, the intelligence community is often required to come to some judgment as to whether enemy preparations are to be considered offensive or defensive in nature—in short, does fear of attack by us or some other adversary account for his military activity, or is he engaged in preparations for offensive action? We will be considering this problem in later chapters and in other contexts. Some comment is also pertinent here, however, because there is often a tendency to write off many enemy preparations as "defensive" when in fact they should be considered as part of an offensive buildup.

This analytical problem is encountered both with respect to buildups of forces which could be used for offensive action (particularly ground force combat units) and to activity which is clearly of a defensive nature (such as increased antiaircraft defenses and fighter patrol activity). There are instances in which the seemingly offensive buildup of ground combat force, and even a considerable one, may be entirely for defensive purposes because of a valid fear of attack. And there are instances in which the seemingly defensive action, the buildup of forces which could never be used in an attack, is in fact part of the offensive preparations and undertaken solely in expectation of retaliation for the impending offense. How can we tell the difference? When should accelerated military activity, of either type, be considered a manifestation of offensive preparations, and when not?

The answer of course is that there is no categorical answer to this question and that each case must be separately considered (this is one reason why warning judgments are so difficult). The following, however, are some general guidelines to be considered:

- Any large-scale buildup of combat strength, particularly major deployments of ground force units, which is in excess of a reasonable defensive requirement for the area, should be considered a probable preparation for offensive action. Experience, in fact, teaches us that it will often be the best single indicator of aggressive intentions.
- Other extensive military preparations (such as a major mobilization effort and a large-scale buildup of logistic support) will reinforce the likelihood that the troop deployment is probably for offensive purposes.
- Preparations which are seemingly "defensive" (particularly air and civil defense measures) can be accurately evaluated only in the light of the buildup of capabilities for offensive action. Where the offensive capability is being rapidly or steadily augmented, the probability is that the concurrent acceleration of "defensive" measures is in fact in preparation for an expected retaliation for the planned offensive.
- In the absence of a buildup of offensive capabilities, the defensive preparations are probably indeed just that.
- The speed and urgency of both offensive and defensive preparations must also be considered. When an offensive action is imminent, the aggressor is likely to be most urgently concerned with the security of his homeland and its populace, and the defensive measures against possible retaliation may be truly extraordinary and greatly accelerated (such as those in mainland China shortly prior to the Chinese intervention in Korea, after a summer in which Peking had shown no great concern with civil defense).

18

Coping with Extraordinary Military Developments

As the discussion thus far hopefully has made clear, the military situation which prevails in the course of preparations for hostilities is extraordinary. It is not a question of degree alone—not just more high command activity, more communications, more alerts, more troop movements, more ammunition shipments, etc. It will also involve activity which is unprecedented, which would never occur except in preparation for or expectation of hostilities. Obviously, the greater the scope of the potential hostilities, the greater the risk to the aggressor of retaliation, and the graver the possibility of escalation (particularly to possible nuclear warfare)—then the greater the deviation from normalcy is likely to be and the greater is the probability that certain measures unique to major hostilities will be undertaken. Clearly, the USSR did not need, for the invasion of Czechoslovakia, to implement many measures which would be necessary or likely in case of an impending nuclear war with the US. (The intelligence problems which might be associated with the possible coming of World War III are discussed in a later chapter.)

The fact that, in every instance, real preparations for hostilities involve activity which is unique to combat does not mean, of course, that we will necessarily detect such preparations. Nor does it necessarily mean that we will recognize such indications for what they are if we do obtain the information. Experience tells us that our capabilities for misjudgment or self-deception in such cases are considerable. (As someone said, no amount of specific evidence would probably have led us to a correct judgment of Chinese intentions in October–November 1950, so great was our misunderstanding.)

Highly Specific Indications

Compilers of indicator lists have devoted much effort to attempting to isolate and to define accurately those preparations which would be unique to hostilities, or at any rate highly unusual in time of peace. Heading any list of such indicators, of course, would be the obtaining by any means (intercepted communications, a high placed agent, captured documents, or otherwise) of the order to attack. Second only to this is probably the plan of attack (without the order to implement it). It may be noted in passing that in the history of warfare such intelligence coups have been accomplished more than once, but not always recognized as such. The Dutch, for example, were correctly apprised of Hitler's plans to attack them in 1940, but failed to accept the information as valid.

Just below such indicators in specificity and value are those which can only be interpreted as preparations for offensive action, which would never be required for defensive or precautionary purposes, or which are rarely if ever undertaken in peacetime. In some cases, such information may be even more valuable or convincing than the plan or order for attack—since the latter will often be suspect, and with cause, as a plant or deception effort.

Among the types of military indicators of such high specific value are:

- The redesignation of administrative military commands as operational commands, or the change from peacetime to wartime organization—such as a redesignation of Soviet military districts or groups of forces as "fronts."
- Widespread activation of wartime headquarters or alternate command posts.
- Release to control of commanders of types of weapons normally held under very strict controls—particularly chemical or nuclear weapons.
- Mine laying in maritime approaches.
- Assignment of a number of interpreters or POW interrogation teams to front-line units.
- Evidence of positive military deception—as opposed to increased security measures.
- Imposition of extraordinary military security measures, particularly new measures not normally noted at all—such as evacuations of military dependents, removal of border populations.
- Greatly intensified reconnaissance by any means, but particularly against likely targets of an initial surprise strike.
- Sudden adoption of extraordinary camouflage or other concealment measures.

When the intelligence system is so fortunate as to obtain indications of such high specificity, it will be most important that they be recognized as such and accorded the weight they deserve, rather than lost in a mass of other incoming data. In some cases, they will be so obvious or definitive that they cannot be ignored, but in others there may be danger that they will not be given due weight for one reason or another. Some illustrations of highly specific indications from crises of recent years, and how they were assessed at the time, may better demonstrate the point:

- The photography showing the first Soviet strategic missile site in Cuba—a development which was obviously definitive and could not be ignored and which immediately overrode all information and supposition to the contrary.
- North Korea's evacuation of civilians from the 38th Parallel in March 1950—an event accorded insufficient weight as evidence that Pyongyang's major buildup of military forces was intended for offensive action.
- Communist China's order of thousands of maps of Korea for its forces in Manchuria in early November 1950—a development also generally given insufficient weight in assessments of Chinese intentions to intervene.
- The emptying of Soviet ammunition dumps in East Germany and Hungary in late July 1968—a development (discussed in Chapter 16) which was scarcely noticed in the mass of other evidence on the Soviet buildup but which, if given more stress, might have helped to convince some who viewed the Soviet buildup as only "more pressure" on Czechoslovakia.

Many other, and probably even more striking, examples could no doubt be drawn from World War II and other conflicts—study of the Pearl Harbor disaster have unearthed a number of valuable specific indications of Japanese intentions which were either mislaid or misevaluated.

The point is that there may be relatively small developments (in terms of the total scope of activity which is under way) which nonetheless are of great importance as indications because they cannot reasonably be explained away or interpreted as defensive or precautionary measures. If one is attempting to compute probabilities by any arithmetic means, such developments will rate very high on the scale. But, in any case, care must be taken to give these items the weight and value they deserve—as the post facto studies of the intelligence performance are sure to do.

Other Extraordinary Developments

In addition to those military developments anticipated on indicator lists which one can expect to occur in the true combat buildup, there will nearly always be some totally unexpected surprises, the things that nobody anticipated. These too are the mark of the exceptional situation. Unfortunately, however, the significance of some of these will not be readily apparent. In some cases, the analyst may be able to obtain assistance from experts on obscure topics or technical questions. In other cases, no one may be found who can interpret the anomaly. It may never be resolved, or the analyst will understand it only after the crisis is past when he will review the evidence for his post-mortem study and suddenly exclaim, "So that's what that meant!" Too late—another lost indication.

The warning or current analyst responsible for analysis in an impending crisis may need to keep up a constant stream of questions to those who might be able to interpret obscure military or technical data for him. No potential source of help should be overlooked, and the channels should be kept open for analysts on all potentially relevant topics to talk with one another. This will be particularly true where much technical material is involved, or information from rarely used sources. Topics never viewed as important by most people before will suddenly be relevant to the flap, and analysts who have never had to work on current subjects will find their services much in demand, if their talents can be brought to bear on the situation. The indications analyst should throw nothing away just because he does not understand it. It may be his missing clue which he will want next week.

The Magnitude of the Buildup

No matter how many unusual or highly specific indications may be noted, assessments in the end will often depend heavily on the sheer magnitude of the military buildup of the potential aggressor. And rightly so. The massive buildup of military power, out of all proportion to anything required in normal times or for defensive purposes, has proven time and again to be the most valid indication of military intent. Many have learned to their regret that they had made a grave error to write off an overwhelming buildup of military force as "just a buildup of capabilities" from which no conclusions can be drawn (see Chapter 5).

Unfortunately, however, these lessons do not necessarily carry over to the next crisis, and the issue must be fought again the next time—and not usually won by the warning analysts either, whose experience does not prevail in these matters. The latest important manifestation of this phenomenon was the invasion of Czechoslovakia, in which many of the same fundamental analytical errors (both military and political) were demonstrated as in the Chinese intervention in Korea. If the most serious misjudgment was essentially a political one (a matter we shall address in the next main section), the military misjudgment was close behind. Essentially, this misjudgment was to regard the massive buildup of Soviet and Warsaw Pact forces as *equally* compatible with an intention to invade and with a mere show of force or attempt to "threaten" the Czechoslovaks. And this despite the fact that the USSR never announced its military buildup, never directly threatened the Czechoslovaks with invasion nor publicly suggested how great its capability was to overwhelm them. Nor was the military buildup all that obvious, either; it was only because our collection capabilities in this area are relatively good and because no extraordinary additional security measures were imposed that we appreciated its scale at all—and then we missed a few divisions. Thus in fact the Soviet buildup was not "saber-rattling" but a bona fide and an astounding buildup of combat force.

This was the largest buildup of Soviet combat power since World War II, exceeding considerably the force which had been brought to bear to suppress the Hungarian revolt in 1956. Experienced military analysts on the Soviet Union had never seen anything like it and were literally stunned by the scale of the effort. "I just can't believe it," and "I never expected to see anything like this except for World War II," were typical analysts' comments to one another. Although the order of battle figures fell somewhat short of what was committed in the invasion, there was nonetheless abundant evidence that forces capable of overwhelming Czechoslovakia had surrounded the country and were clearly deployed along the main lines of advance for an invasion and not engaged in exercises. Sheer logic, not to mention the lessons of history, should teach us that a force of this magnitude was never required just to put more "pressure" on Dubcek. This was a force which met Soviet doctrinal concepts for offensive operations even when resistance might be expected, and there was minimal sign indeed that the Czechoslovaks were likely to put up much resistance.

How much evidence do we need to make a judgment that a military buildup of such proportions reflects a probable intention to attack? Some analysts feel that if this judgment could not be made by a majority on the basis of the evidence which was at hand on Czechoslovakia, the majority can probably never be expected to make such a judgment. Yet it will be the judgment of history, and the experience of warning analysts, that buildups of this magnitude carry a high probability of offensive action. Which leads us to—

Redundancy

It is a feature of both the buildup of combat forces and of logistic preparations for war that the aggressor will seek a great superiority of power and ability to sustain combat in comparison with that of his adversary—that is, if the military planner gets his way. All commanders want an abundance of everything before they attack. They are practically never satisfied that they have enough. Occasionally, military forces will have to operate on a shoestring, but rarely at the choice of their commanders.

Thus a redundancy of supplies and a great superiority in equipment, as well as massive troop deployments, may be the mark of the true buildup of military forces for attack. The enemy, if he has the capability and the time to do so, will sometimes appear to be inordinately long in his buildup and to be undertaking all kinds of preparations which may seem to us to be unnecessary or at least much more than is necessary.

The intelligence analyst in these circumstances should take care to avoid the pitfall into which more than one of them has fallen, which is to conclude that all this buildup does not mean anything because the enemy already has more than enough on hand to attack. This argument has been seriously advanced, and even propounded as a negative indication ("he would not do all this unless he was bluffing"), on several occasions in the past twenty years or so. The ensuing action was likely to be an overwhelming military action (e.g., the invasion of Czechoslovakia). This kind of inverse reasoning is the death of warning and the mark of an analyst who will find any excuse to explain away what the enemy is up to.

Massive buildups of combat power and logistic support are *never* negative indications. They may occasionally (repeat, occasionally) not be followed by military action, but the odds are high that such action will follow. Excessive military preparations are not bluff nor meaningless just because you think that the adversary has enough on hand already. As a general rule, the greater the ratio of buildup in relation to that of the opponent, the greater the likelihood of attack. It is fatal to ignore evidence of this kind.

19

Importance of Political Factors for Warning

It is easy to demonstrate the importance and relevance of military developments for warning. Anyone can recognize that the numerous military preparedness steps identified on indicator lists bear a direct relationship to a capability, and hence at least a possible intention, to commit military forces. Many military developments, including some of the most important, are physically measurable or quantifiable—assuming, of course, that the collection capability exists. There *are* so many tanks deployed in this area, which represents such-and-such a percentage increase over the past two weeks, etc. Such information, factually speaking, is unambiguous; its interpretation is not dependent on subjective judgment.

Ambiguity of Political Indicators

In contrast, the relevance of political developments or political indicators to warning is often not so readily apparent, is not factually demonstrable, and interpretation of specific developments is likely to be highly subjective. The potential for concealment of intention in the political field, not to mention for deception, is much higher than for military preparations. At least in theory, it is possible for a closed society to conceal completely its decisions, to fail to take measures to prepare its own populace psychologically for war, and to handle its diplomacy and manipulate its propaganda so that there is virtually no discernible outward change in the political atmosphere which might alert the adversary. In practice, of course, this virtually never occurs.

But, even when there are numerous political anomalies and significant changes in diplomacy and propaganda, the interpretation of their significance may be difficult and elusive. Short of old-fashioned ultimatums and declarations of war, or the collection pipeline into the adversary's decision-making councils, nearly all political indications are subject to some degree of ambiguity and uncertainty. It follows, of course, that interpretations of political indications are likely to be much more variable and controversial than of military developments.

One manifestation of this is that there are usually fewer political and civil developments on indicator lists and that they tend to be much vaguer and imprecise in wording. An illustration or two will suffice to make this point. Indicator lists usually carry such political items as: "Protracted high-level leadership meetings," and "Marked intensification of internal police controls." Such developments are, of course, potentially significant indications that decisions on war are under consideration or have been taken, but they may also be attributable entirely to domestic developments, such as civil unrest. Even political indicators tied directly to foreign affairs, such as "A general hardening of foreign policy" or "Significant increases in propaganda broadcasts to or about a critical area," are not in themselves necessarily manifestations of any decision or intention to resort to conflict. Such developments are significant even as possible indications only in relation to what is sometimes called "the overall situation." Although not all political indicators are so unspecific, it has not been possible to define potential political indicators with anything like the precision which is possible for military developments; there is no political TO&E. Nor is it possible to forecast in advance whether an adversary will choose to publicize his objectives and intentions, seek to conceal them almost totally, or—as is most probable—take some intermediate course. Thus the number of political indications versus military is almost impossible to anticipate for any hypothetical future situation. We can forecast with some degree of confidence that some specific military preparations will be undertaken, but we cannot forecast or at least cannot agree what manifestations of the political decision may be evident, or how such manifestations should be interpreted.

This ambiguity and non-specificity of political indicators also often means that our sense of "political warning" is likely to be much more subjective, and hence more difficult to define or explain to others than is the military evidence. Sometimes, there is little more than an uneasy sense or intuitive "feeling" that the enemy is up to something, which of course is not provable or even necessarily communicable to others who are not thinking on the same wave length. The analyst or military commander which attempts to put this sense of unease into words may feel almost helpless to explain his "feelings," if not downright apprehensive that he is making a fool of himself. Yet, often these "feelings" have been generally accurate, if not specific, barometers of impending developments. Thus, General Lucius Clay, a few weeks before the start of the Berlin blockade, dispatched a cable to Army Intelligence in Washington, which said in part "Within the last few weeks, I have felt a subtle change in Soviet attitude which I cannot define but which now gives me a feeling that it [war] may come with dramatic suddenness. I cannot support this change in my own thinking with any data or outward evidence in relationships other than to describe it as a feeling of a new tenseness in every Soviet individual with whom we have official relations."[1] Or, as General Clay later recalled his feelings at the time: "Somehow I felt instinctively that a definite change in the attitude of the Russians in Berlin had occurred and that something was about to happen. I pointed out that I had no confirming intelligence of a positive nature."[2]

Much the same sense of unease that something was about to happen has haunted perceptive intelligence analysts on other occasions—such as the spring of 1950 prior to the North Korean attack in South Korea, and in the early months of 1962, even before the marked upsurge of Soviet shipments to Cuba was begun.

Even when political warning is less vague and subjective—that is, when the political atmosphere is clearly deteriorating and tensions are rising over a specific situation which may lead to war—the political indicators still may be imprecise and not measurable or quantifiable evidence of a specific course of enemy action. There are, of course, exceptions to this, in which the adversary may make no attempt to conceal his plans, or in which direct warnings are issued to the intended victim, both privately and publicly. Often, however, political indications can give us only generalized warning, such as a recognition that the dangers of war are increasing substantially, or that the enemy is clearly committed to some course of action which entails a grave risk of hostilities.

Critical Role of Political Factors for Warning

In large part because of these uncertainties, there exists a fundamental mistrust and misunderstanding of the importance of political factors for warning. Particularly among military officers, although by no means confined to them, there has been a tendency to downplay the significance of "political warning." Because political indications are less precise, less measurable and less predictable than military indications, it is an easy step to conclude that they somehow matter less, and that we can give them only secondary or incidental attention in our assessment of the enemy's intentions. Warning papers and estimates, in some cases, have seemed to place undue emphasis on the detection of military preparations, with only passing reference to the political problem. The importance of the political assessment has rarely been so well defined as in the following perceptive comments written several years ago by one of the few real warning, experts in the US intelligence community:

> We query whether the critical role of the political factors in warning may not warrant somewhat more emphasis or highlighting.
>
> We appreciate that, in the warning process, the political factor or "posture"—constitutes one of the most esoteric and elusive fields. The very term "political posture" remains essentially ambiguous. Nevertheless, elusive, ambiguous or no, its critical role in warning must be duly weighed. We discern the implication that the political factor somehow constitutes a separable category, distinct from that of physical preparations; that when joined with the latter at some point well along in the game, it may, mathematically, add to or subtract from, the sum of our physical holdings. Actually, the political context is determinative of whether at any and every given point in the progress of the enemy's preparations you indeed hold any sum at all in the "preparations" category. The political context to us is not merely another increment to the warning conveyed by a particular pattern or patterns of observed physical "preparations." It is rather the essential, *a priori* context which establishes that a particular physical activity may have any possible relevance to a real, live warning issue; it gives or denies to the physical "preparations" their presumed evidential value as indications. In any discussion of a hypothetical future warning problem, there is, of necessity, present an exquisitely subtle subjective assumption. Any discourse on what indications or evidence one expects to receive, how one will handle these, etc., assumes the very point at issue—that one is dealing with activities recognized as "preparations." Now, logically, there cannot be "preparations" for something that in fact the enemy has no conscious design of doing; there cannot be a valid "indication" of that which does not exist in reality (much less a whole compendium of such "indications" sampled from a cross-section of the attack-bound enemy national entity). Unless and until, and then only to the degree that, the intelligence community's intellectual assumptions and convictions as to the enemy's political posture can rationally accommodate at least the *possibility* that the enemy just *might really* be preparing to attack, there is not likely to be acceptance—even contingent and tentative—of any enemy activity whatever—specific or in pattern—as reflecting or indicating "preparations" to attack. So long as the prevailing political assessment of the enemy's foreign policy objectives, motivational factors, etc., confidently holds that the course of action for which alleged "preparations" are being made is inconceivable, or impossible (or even unlikely), there

has not even been a beginning of the cumulative process [of indications intelligence]. Thus the political factor invariably stands athwart the warning exercise from the very outset, and represents a constant, vital ingredient in the warning process from beginning to end.

Our remarks above derive primarily from our cumulative experience in the warning process. The same conclusion, however, follows from the intrinsic logic of the problem itself. The very end to which warning generally addresses itself—enemy intent to attack—is fundamentally a political issue, involving a political decision of the highest order, made by the political leadership of the enemy state (we are excluding here, of course, the "Failsafe" issue of some military nut just arbitrarily pushing a button). The working rationale underlying the exercise discussed throughout this estimate is simply the presumed existence of:

(a) an enemy decision to attack;
(b) a plan of measures/preparations to be taken to ensure success of the attack;
(c) implementation of the plan.

The intelligence processes involve basically our attempt to detect, identify, and place in order fragmentary manifestations of the process actually under way in (c), with a view toward reconstructing and authenticating the essential outlines of (b), from which we hope to derive and prove (a) which equals classic warning. We cannot hope to reason effectively from (c) to (a) without a correct, albeit hypothetical, appreciation from the outset of (a). Here again, then, we find that in theory as well as in practice the crucial, final link is entirely political. Whether viewed from Moscow or Washington, the political context is the capstone: for the enemy—the beginning of that fateful course; for US intelligence—the end!

Political Perception Fundamental to Warning

The perception of the enemy's fundamental goals and priorities is the *sine qua non* of warning. It constitutes the most significant difference between those who "have warning" and those who do not. No amount of military evidence will serve to convince those who do not have this political perception of the adversary's objectives and national priorities, or those who cannot perceive that military action may be the rational outcome of the adversary's course of action to date. The validity of this point can be demonstrated in instance after instance; it is the problem of "those who cannot see," and more "facts" will have little effect on their ability to see. Just as some could not see that Hitler was bent on conquest in Europe, others later could not see that China would or even might intervene in Korea or that the USSR would or even might invade Czechoslovakia. All were fundamentally problems in political perception, rather than the evaluation of military evidence. An indications study on the Czechoslovak crisis (written after the event) described the analytic problem as it existed in mid-July (just prior to the start of Soviet mobilization and major troop deployments) as follows:

It is important to note that, while current intelligence reporting at this time clearly and explicitly recognized the gravity of the crisis and the nature of Soviet tactics, there was also a fundamental difference of opinion among analysts. The point at issue was the means which the Soviet Union could and would use to accomplish its objectives and whether it would, if faced by continuing Czechoslovak intransigence, ultimately resort to overt intervention in Czechoslovak affairs.

On the one hand, there was a group of analysts who questioned whether there was anything that the Soviet Union really could do, including employment of military force, to reverse the trends in Czechoslovakia. This group was also inclined to the view that the USSR, if unable to secure Czech compliance by political means, would not jeopardize its international image, its relations with western Communist Parties and its progress toward coexistence with the United States by direct military action. It believed that the USSR had changed or matured politically since the days of the Hungarian intervention and was unlikely to take such action again. For these reasons, direct Soviet action against Czechoslovakia was viewed as somewhat "irrational" and therefore unlikely. This group was thus predisposed, in varying degrees, to regard subsequent major Soviet military moves as more pressure on Czechoslovakia rather than as bona fide preparations for military action.

On the other hand was a group of analysts who inclined to the belief quite early in the summer that the USSR was deadly serious in its determination to maintain control of Czechoslovakia and would ultimately use any means 1 including military force, to ensure this. They believed that the USSR, not just for political but also for strategic reasons, could not tolerate the loss of Czechoslovakia and that Soviet security interests were the paramount consideration. The USSR therefore would decide, if in fact it had not already decided that military action against Czechoslovakia was the lesser of the evils which confronted it. These analysts thus did not regard such a course of action as irrational and they were predisposed earlier rather than later: to regard the Soviet military moves as preparations for direct intervention.

Such judgments or estimates by individuals are crucial to the warning process, and each person makes his own regardless of whether there is an agreed national estimate. Each analyst is influenced, perhaps unconsciously, by his preconceived views or his opinion of what is rational or logical behavior on the part of the enemy. His judgment on this will help to determine, sometimes more than he may realize, not only how he interprets a given piece of information but what he selects to report at all.

The foregoing discussion should help to explain why some critics object to the terms "military warning" and "political warning" as if they were separate processes. There are indications which are essentially military, and those which are primarily political, but there is only one kind of warning. It is the perception of the significance of all these developments in toto. Warning, like beauty, lies in the eye of the beholder.

It is highly erroneous to presume, as many do, that political analysts (or even political agencies, such as the Department of State) make political analyses. and that military analysts (and military agencies, such as the Department of Defense) make military analyses. The intelligence offices of the Department of State do a great deal of essentially military analysis and must constantly take military factors into account in making political assessments. Still more pertinent, perhaps, is the fact that military analysts are constantly making essentially *political* judgments about the likely *military* courses of actions of our potential enemies. They may not recognize that this is so; it may be entirely unconscious, but assessments of political factors underlie virtually all military estimates and other analyses of enemy courses of action.

It may be extremely important for warning that gratuitous political judgments of intent do not creep into military assessments of the enemy's capabilities, or at least that the political judgment be clearly separated from the statement of the military capability. This point was well illustrated in judgments made prior to the outbreak of the Korean War in June 1950. By March, Army Intelligence was correctly reporting that the steady buildup of North Korean forces gave them the capability to attack South Korea at any time—but then undercut this significant military judgment (a warning in itself) with the judgment that it was not believed that North Korea would do it, at least for the time being. The same positive military but negative political judgments were being made by General MacArthur's intelligence in the Far East. Thus, the gratuitous political judgment (the basis and argumentation for which was never really set forth) tended in effect to dilute or even negate a highly important military estimate.

Warning has failed more often for lack of political perception than it has for lack of military evidence. When I have pointed this out to military officers, their reaction often is that you really cannot trust these political people, and if more heed had been paid to the military people, all would have gone well. In some cases this may be true (it is at least partly true for Czechoslovakia in 1968), but in other cases it has been the military themselves who have permitted their own political misperceptions to override the military evidence. And there are cases in which political officers have been well ahead of the military analysts in perception of the likely course of military action. In warning, where we have so often been wrong, there is blame to go around. The point is that the political judgment, no matter who makes it, will likely be even more important than the military analysis for the assessment of the enemy's intentions.

Notes

1. Walter Millis, ed., *The Forrestal Diaries* (New York, The Viking Press, 1951), 387.
2. Lucius D. Clay, *Decision in Germany*. (New York, Doubleday & Co., Inc., 1950), 354.

20

Basic Political Warning— A Problem of Perception

PRESUMABLY, FEW PERSONS would take exception to the general thesis set forth in the preceding chapter—that the perception of enemy intentions is essentially a political judgment of what he will likely do in given circumstances, and that this understanding of national priorities and objectives is fundamental to any warning judgment. In this chapter, this problem will be examined in more detail, with some specific illustrations of how it may affect our appreciation of the enemy's course of action.

Perceptions of the Likelihood of Conflict

In normal circumstances, the likelihood or unlikelihood of conflict between two or more nations is fairly well understood, not only in government circles but by the educated public, and these judgments are usually quite accurate. They are derived from our recognition that the basic conditions for war between two or more nations either do not exist at all, or are present in varying degrees of probability ranging from a very small chance that war would occur, to the situation in which virtually all political signs of ultimate hostilities are positive.

We may cite two examples of these extremes. War between the United States and Canada, given the present international situation and particularly the political systems of the two nations and their long tradition of friendship, appears virtually inconceivable. For war to occur between them, there would have to be drastic and fundamental political changes, which would be clearly recognizable. At the other extreme, the formation of the state of Israel, in the midst of Arab countries which are bitterly opposed to it and indeed deny its right to exist at all, has created a situation in which the fundamental conditions for war are present at all times—and which have led to three wars in 20 years. Until or unless the political atmosphere is drastically changed, the probabilities are that war between Israel and the Arab states will erupt at some time again. There need be no fundamental worsening of the atmosphere for this to occur. Politically speaking, war is always a probability, and it is largely military restraints, particularly the superiority of Israeli military forces, that thus far have served to prevent its recurrence since the 1967 conflict.

Between these two extremes, there are all sorts of gradations of our assessments of the probabilities of conflict between nations, or of our perceptions of whether a given nation is or is not inclined to resort to hostilities, or of how aggressive or cautious it may be in pursuit of its aims. Our understandings of these questions are basic not only to our intelligence assessments (particularly our national intelligence estimates) but also to our national political and military policies. Our views concerning whether the USSR would or would not launch a surprise military attack on the US if it felt it could get away with it, or whether China would back another attack by North Korea if it felt the risks were tolerable, are absolutely fundamental. They determine in large part not only how we assess particular moves by these nations, but more importantly what risks we think that we can run and how much of the national budget is allocated to defense against the contingency of attack by our potential enemies.

A brief examination of our changing attitudes toward the Soviet Union and China since World War II will amply illustrate how fundamental these attitudes are both to our national policies and to intelligence assessments of their likely courses of action.

Our military alliance with the USSR in World War II, together with the US desire to end the war as soon as possible and bring the boys home, was conducive to an atmosphere, or euphoria, in which suggestions that the USSR might have some aggressive post-war ambitions were initially most unpopular, and often completely rejected. The writer recalls how difficult it was to gain a hearing in Army Intelligence in 1944–45 for the idea that the USSR had subversive designs in Latin America. It took a series of unpleasant shocks—the repression of Eastern Europe by Soviet forces, the Soviet invasion of Iran in 1946 and threat against Turkey in the same year, Communist guerrilla warfare in Greece, and the Communist subversive threat to Western Europe—to bring about a major change in national attitudes, policies, and intelligence assessments of Soviet intentions. There followed, in 1948, the

Communist coup in Czechoslovakia and the Berlin blockade, and in 1950 the Soviet backed North Korean attack on South Korea. This series of cold war crises with the Soviets ultimately brought about an atmosphere in which many in this country, including some in the intelligence field, became convinced that the USSR was seeking world domination by force, and that it was deterred from further military ventures only by US strength and willingness to respond to the threat. Yet, in little more than a decade (following the Soviet retreat in the Cuban missile crisis and the ensuing relaxation of US-Soviet tensions), attitudes had swung so far back the other way that many had come to view a detente with the US as the highest priority in Soviet policy. To these people, the Soviet invasion of Czechoslovakia came as a rude surprise—not for lack of warning, of which there was ample, but for lack of perception of Soviet objectives and priorities.

The China problem over the same period has presented a not dissimilar picture. A view, widely held in the 1950's because of the intervention in Korea and the threat of intervention in Southeast Asia, was that the Chinese Communists were highly aggressive and intent on conquest of much of Asia. Today, almost the opposite view has become accepted—that they are highly preoccupied with building up their country and would enter into foreign military ventures only if they saw their own security to be directly threatened. Needless to say, these changing attitudes have tended materially to affect our assessments of possible indications of their military intentions.

Such attitudes are crucial to our views as to the nature, and amount, of "political warning" which we may receive. If the opinion prevails that the national leadership of any country is essentially aggressive and bent on expansion or conquest, then it follows that there need be no basic political changes in the attitudes or behavior of that country before it attacks. In this case, the *casus belli* already exists; it is not brought about by some change in circumstance which we will be able to perceive.

This problem of how much "political warning" we would receive of Soviet attack in Europe has been a bone of contention in NATO planning and estimates, with national positions often closely keyed to the proximity of the potential enemy. Thus, those nations of Western Europe most immediately in danger of being overwhelmed by a Soviet ground force attack have often placed less faith in "political warning time" than have those, such as ourselves, who are not in such immediate danger. Most US analysts today would agree that the outbreak of war in Europe is now almost inconceivable without a fundamental *and discernible* change in the political situation, which would generate a period of political tension in which the growing possibility of war would be evident. Nonetheless, there remain some who do not accept this even today, just as—at the peak of the cold war in the late 1940's and 1950's—many analysts would have questioned that any specific "political warning" of Soviet military intentions would have been received. There have been many military papers, some of quite recent vintage, which have maintained that the USSR not only *could* attack in Europe with little or no warning, but probably *would* do so. Such judgments are both military and political; they assume both that there would be no significant military buildup or other discernible preparedness measures prior to the attack, and also that there would be no major change in the political atmosphere which would raise the likelihood of conflict. Obviously, our assumptions and attitudes on this question are absolutely vital to our views on warning of attack in Europe, and the opposing views have proved virtually irreconcilable.

Long-Term Factors: Priorities and Traditional Behavior

How any nation may react in a particular situation will usually be predicated, at least in part, on its traditional national objectives and past performance. It is thus essential to understand what the national objectives or priorities of the potential enemy have been in the past. This, of course, assumes a rationality and consistency of national behavior which have not always been the case, but the premise nonetheless is usually valid. It is from such concepts of likely national behavior in certain circumstances that we derive our judgments of what a nation will fight for and what it will not. We know from experience that the Soviet Union has fought (or at least used its military forces) to preserve its hegemony over its empire in Eastern Europe, and we therefore tend to deem it likely that it would, if necessary, do so again. On the other hand, experience thus far has taught us that the USSR is not likely to run the risks of fighting to gain new territory.

It is virtually impossible for any country (or leader) to conceal for long what its basic philosophies and national objectives are. Deception and concealment cannot extend this far. All leaders need some popular support for their programs, particularly programs which may ultimately lead to war. History shows that most leaders, even those bent on a course of aggression, rarely have made much effort to conceal their intentions, and some leaders (e.g., Hitler in *Mein Kampf*) have provided us with virtual blueprints of what they planned to accomplish. If we do not, in such cases, correctly perceive the enemy's general course of action, it is often because we did not wish to believe what we were being told—just as many refused to accept the clear warnings from Hitler's own writings.

Unfortunately, however, political warning is not this simple. While it is essential to understand the potential enemy's fundamental objectives and priorities, this is not likely to provide us with specific warning of what he will do in some particular situation. It usually cannot tell us how great a risk he is prepared to run to achieve his objectives, how far he may seek them by political means before he will resort to military action or whether he will, in fact, ever finally take the military course. In short, even when our understanding of the adversary's philosophy and objectives is pretty good, we must still have some more specific understanding of his objectives and decisions in the specific situation in order to predict his likely course of action.

Strategic Importance of the Particular Issue

Except in instances of long-planned deliberate aggression, the possibility of conflict usually arises over some particular issue or development, and the potential aggressor may have had very little control of it (see discussion in Chapter 26). Or, if the situation is largely of his own making, development of the situation and the reactions of others may be different from what he had expected. There are potentially, and often actually, a vast number of complicating factors which may influence his political decisions. It will not be enough just to have a general estimate of how he should react in such circumstances, or how he has reacted in the past. It is important to understand how he views the situation now and to interpret how he will behave in this particular instance. We are confronted now with a condition and not a theory. We thus move from the long-term estimative approach to the problem to the specific and more short-term indications approach.

How much weight are we going to give, in these circumstances, to our traditional concepts of this nation's objectives and likely courses of action, and how much to the specific indications of what he is going to do this time? In a fair number of cases, there is not apt to be a great deal of conflict here—the traditional or seemingly logical course of action will in fact prove to be the right one. In this case, the current political indications will be generally consistent with how, we expect this particular nation to perform. This will be particularly true if both past behavior and current indications call for an essentially negative assessment—i.e., that the nation in question will not resort to military action in these circumstances.

The difficulties in warning are likely to arise when some of these factors are out of consonance with one another, and particularly when standing estimates or judgments would dictate that the adversary will not take military action in this situation, but the current indications, both military and political, suggest that he will. Which is right, and what validity should be given our current indications as against the going estimate?

Any answer without numerous caveats is likely to be an oversimplification and subject to rebuttal with examples which will tend to negate the general conclusion. History nonetheless suggests that the greater weight in these cases should be given to the current indications. In other words, it is usually more important to understand the strategic importance of the particular issue to the nation than it is to place undue weight on traditional behavior and priorities, This is, after all, the fundamental cause of warning failures—that the behavior of the aggressor appeared inconsistent with what he would normally have expected him to do, or with our estimate of what he would do. Thus, we were "surprised." He did not do what we thought he would do, or should do.

In some instances, the enemy's course of action truly does appear irrational. It is a misjudgment of the situation, in either the long or short term, or both, and in the end it is counterproductive to him. Two conspicuous examples which come to mind are Pearl Harbor—which was a short-term triumph but long-term misjudgment on the part of Japan—and the Cuban missile crisis, which was a gross miscalculation in the short term. In both, the indications of what the enemy was doing were more important to an assessment of his intentions than any going estimates, which in fact proved to be wrong. At was observed, in one of the numerous post-mortems on the Cuba crisis, that we had totally misjudged Khrushchev's sense of priorities (just as he had misjudged ours) and that there must have been an overriding requirement in his mind to achieve some degree of strategic parity with the US which would have led him to take such a risk.

In lesser degree, this may be said of many crises. The perception of what the enemy is thinking and how important the current issue is to him is fundamental to our ability to understand what he will do. It was a lack of such perception that lay behind much of our misjudgment of North Vietnamese intentions and persistence in the Vietnam war. As has subsequently become obvious, both US intelligence and perhaps to a greater degree policy levels (there were individual exceptions, of course) vastly underestimated the determination and ability of the North Vietnamese leadership to sustain the war effort. No doubt this attitude contributed materially to the reluctance to believe in 1965–66 that Hanoi was mobilizing its armed forces for the conduct of a prolonged war in the South (see discussion in Chapter 15).

We should note here also that it may require no particular collection effort or sophisticated analytic talent to perceive how nations feel about particular issues. Even our security-conscious adversaries whom we characteristically suspect of all kinds of chicanery are not necessarily engaged in devious efforts to conceal how they feel on great problems and issues vital to their national security or objectives. It will often be quite obvious—how they feel about something and how important it is to them—if we will only take the time to examine what they are saying and try to see it from their viewpoint. In some cases—such as China's intervention in the Korean war and North Vietnam's general program for the conduct of the war in South Vietnam—they have virtually told us what they intended to do. In such as the Soviet invasion of Czechoslovakia, they have made no secret whatever of the criticality of the issue and of its overriding importance to them, and have strongly indicated that force would be used if needed.

Czechoslovakia as a Problem in Political Warning

The Czechoslovak problem is worth some elaboration here, since it provides an unusually clear and interesting example of the nature of political warning, the problems involved, and the lack of understanding of what political warning is. It is hoped that the reader will not tire in the coming chapters of the use of the invasion of Czechoslovakia as an example, but we have had few cases which have served to demonstrate so many valid principles of warning. It is indeed a classic textbook case.

Some weeks after the invasion, this writer was asked to review a draft paper written in one of the military intelligence agencies (which shall go unnamed) concerning the invasion. Among other things, this paper stated that, as that agency had previously maintained would be the case in event of Soviet military aggression, "there was no political warning of the invasion" (sic!). No political warning! We had had political warning all summer long, in repeated and progressive manifestations, for at least five months before the invasion, of the Soviet Union's deep concern with the potential consequences of the liberalization trend. As the "Prague Spring" continued to flourish, the USSR's anxiety over the situation and political and military pressures on the Dubcek regime became ever more evident. At no time did the USSR attempt either publicly or privately to disguise its concern that political developments in Czechoslovakia posed an ultimate if not immediate threat to the political and military hegemony of the Soviet Union in Eastern Europe. So evident was the USSR's preoccupation with this problem that there was some tendency to believe that it had exaggerated the threat and was "overreacting." The Soviet leadership clearly was obsessed with the problem of Czechoslovakia.

There were, moreover, specific warnings of the Soviet Union's intention to invade if political measures failed to achieve the desired goal. Both the French and Italian Communist Parties were so alarmed in mid-July (five weeks before the invasion) that their leaders hurried to Moscow to plead with the Soviets not to invade. Both returned empty-handed and with no assurances from Moscow that military force would not be used. Concurrent with this, the five invading powers convened in Warsaw and issued a virtual ultimatum to Czechoslovakia. Two weeks later, agreements reached at Cierna and Bratislava called on the Dubcek regime to take certain measures to redress the situation, with an implicit threat that failure to comply could lead to military action by the massive forces which now surrounded Czechoslovakia.

What can be meant then by "no political warning"? Presumably, only that the USSR never issued a direct public threat to use military force, and that there was no last-minute ultimatum or dramatic shift in propaganda which would have signaled that invasion finally was imminent. But rarely has basic political warning of intention been more evident. The USSR never attempted to conceal its intention to bring the Czechoslovak situation under control and in fact tried virtually every other device at its command before it finally resorted to military force. To maintain that political warning was lacking in this circumstance is to misunderstand political warning. While sometimes it may be explicit and specific, it often will be generalized. It usually will not provide us definite evidence that final decisions have been reached and particularly that military action is imminent, since most nations will seek to withhold such information in the interests of achieving tactical surprise (see Chapter 28).

The NATO powers, which were largely surprised by the invasion of Czechoslovakia, in retrospect came to recognize that there had in fact been ample warning. A post-mortem read: "We—and the Czechs—had several months of quite visible political warning, plus a number of weeks of strategic warning, as Warsaw Pact forces got into position to threaten the Czech leaders with a military invasion."

Some Factors Influencing Political Perception

We have addressed in other chapters the human factor of perceiving and believing evidence which is in conflict with one's preconceptions, and we will be returning to this subject again in later chapters. Objective perception of the enemy's attitudes and the ability to look at things from his point of view are crucial to warning, and above all to political analysis, since this will necessarily be more subjective than the compiling and analysis of military data.

The "climate of opinion" also strongly influences political perception, as noted earlier in this chapter with regard to our changing national attitudes toward the USSR and China since World War II. It is not only very difficult for an individual to maintain an independent viewpoint against a widespread contrary view about another nation, it may prove almost impossible to gain acceptance for such a view, even when there may be considerable evidence to support it. Time is needed to change national attitudes.

A somewhat related factor may be the influence of our own national policies and military plans on our judgments of what the adversary may do. Once a national decision has been made on a certain course of action—such as whether a particular country is or is not vital to our defense and hence whether we will or will not defend it—there will almost inevitably be some impact on our assessment of that nation's actions. It is not so obvious as simply saying what the policy level would like to hear (or not saying what it would like not to hear); there tends also to be a more subtle influence on our thinking and analyses. Various historical examples could probably be cited, Vietnam for one. Our concepts of North Vietnam as an aggressive nation bent on conquering the South almost certainly were influenced, or at least reinforced, by the US decisions in 1965 to commit forces to defend the South; it then became acceptable to talk of North Vietnam as an aggressor and hence to think in such terms.

Judgments concerning North Korean intentions in the period prior to the attack of June 1950 also were materially influenced by US policies in that area. For at least three years before that attack, it had been officially recognized that there was grave danger that North Korea would seek to take over the South if US forces were withdrawn. Nonetheless, it was decided to withdraw US forces, partly on the grounds that South Korea was not essential to the US military position in the Far East, and to hand the Korean problem to the United Nations. Once having decided to write off South Korea as a US military responsibility, the US made no military plans for the defense of South Korea against an attack from the North, and seemingly it became US policy not to defend South Korea. The effect of this on intelligence assessments, and thus indirectly, on warning, was two-fold: as a low priority area for US policy, Korea became a low priority collection target; and intelligence analysts, believing that the US would take no military action if North Korea attacked, tended to downplay both the importance and by implication the likelihood of the attack in their assessments. Even those who expected the attack and predicted that it was coming (not necessarily in June, of course), saw the possibility as a relatively unimportant development in comparison with other potential Communist military threats in Europe and the Far East, and hence gave it little attention in their assessments. They saw no urgency in warning the policy maker about Korea, since nothing was going to be done about it anyway. It was only one of many areas where the so-called Communist Bloc (meaning the Soviet Union and its obedient satellites) might strike, and apparently one of the least important.

A related factor influencing assessments on Korea in that period was the concept that only the Soviet Union was a real military threat against which US military forces should be prepared to act. The concept of limited "wars of liberation" or indirect aggression through third parties was vaguely perceived, if at all. North Korea, like Communist Europe, was seen only as a pawn of Moscow; war, if it came, would be on Soviet instigation and part of a much larger conflict. Intelligence assessments, as well as military planning, reflected this view of the Communist threat and scarcely hinted at the possibility of a Communist attack which would be confined to the Korean peninsula. General Ridgway has well described the then prevailing concept as follows:

> By 1949, we were completely committed to the theory that the next war involving the United States would be a global war, in which Korea would be of relatively minor importance and, in any event indefensible. All our planning, all our official statements, all our military decisions derived essentially from this belief.[1]

Finally, we may note the effects on judgments of the likelihood of attack of the unwillingness to believe it or to accept it—the tendency to push the problem aside as too unpleasant to think about, in the hope that it may just go away. We have noted this human tendency earlier, and we will be coming back to it in a later chapter, as one of the major factors affecting the warning judgment. This tendency, which all of us have in some degree, may be accentuated by a sense of hopelessness and inability to do anything about it, or by a desire not to rock the boat or stir the waters lest the potential aggressor be even more provoked. This last consideration possibly was a major factor in Stalin's apparent failure to have anticipated the German attack on the Soviet Union in June 1941, and his seeming dismissal of the numerous warnings of the coming attack. There is no doubt that the USSR had ample long-term strategic warning of the German offensive, and some observers have felt that Stalin was blind to this, suffering from a megalomania almost as great as Hitler's. But an alternate thesis holds that he did foresee the attack but, believing that nothing further could be done to prevent it, he sought to delay it as long as possible by trying to appease Hitler and thus publicly refusing to concede that there was danger of attack. Whether true or not—we shall probably never know what Stalin really thought—the effect of his policies was to decrease the preparedness of the Soviet public and particularly the armed forces for the attack when it finally came.

Note

1. Matthew B. Ridgway, *The Korean War* (Garden City, N.Y., Doubleday & Co., Inc., 1967), 11.

21
Some Specific Factors in Political Warning

In this chapter, we shall examine some of the types of political developments which may provide us warning and attempt some assessment of their value as indications. It is very difficult to be specific on these subjects, that is to predict what kinds of political developments are most likely to be of value in any warning situation which may arise in the future, or even whether they are likely to occur at all. Therefore, this discussion perforce is rather generalized and probably inadequate, but will include some specific illustrations from the past.

Diplomacy and Foreign Policy

Since war is a carrying out of political relations by other means and nations will usually resort to war only when they have failed to secure their objectives by political means, the conduct of foreign policy and diplomacy obviously are highly important indications of national objectives. It is difficult to conceive of hostilities breaking out between nations today without some prior crisis or at least deterioration in their diplomatic relations. Indeed, historically, the most obvious early warning of approaching hostilities has usually been in the field of foreign political relations. The outbreak of both world wars in Europe was preceded by marked deterioration in the international political climate, which made the threat of war apparent to all, if not "inevitable." Even Japan's surprise attack on Pearl Harbor was preceded by a crisis in US-Japanese political relations, which had greatly raised US fears of war, although specific

Japanese intentions were not foreseen. Those who are confident that the coming of future wars also will be foreshadowed by international political crises and developments in the conduct of foreign policy unquestionably have the lessons of history on their side.

Nonetheless, there is a substantial body of opinion which questions the likelihood that wars of the future will necessarily be preceded by such obvious changes in the political atmosphere. Moreover, it is our uncertainty that political indications of this nature will provide us warning that largely accounts for the existence of indications intelligence at all. If we could be confident of this type of political warning, not to mention old-fashioned ultimatums and declarations of war, then obviously there would be little need for much indications analysis. We could confine ourselves to assessments of the enemy's capabilities.

The circumstances surrounding the outbreak of some conflicts since World War II certainly justify this concern. The North Korean attack on South Korea, the most conspicuous example, was not preceded by any political crisis or diplomatic warning in the near term, although the political atmosphere had long been highly strained, and of course the two sections of the country had no diplomatic relations. The diplomatic warnings of Chinese intervention in Korea—although they were issued—fell short of what might have been expected if the Chinese objective was truly to deter the advance of US/UN forces toward the Yalu. The Middle East conflicts of 1956 and 1967 were both preceded by international political crises, but specific political indications that Israel had decided to attack were largely lacking.

There are perhaps three major reasons that we have less confidence that we will receive specific political warning through developments in foreign policy and diplomacy than has been true in the past:

- Modern weapons, even non-nuclear weapons, have given a greater advantage to the attacker, thus increasing the value of political surprise. The Israeli attack of 1967 is a prime example. Probably in part because of this, it is no longer considered desirable to break political relations or to declare war prior to attacking, and few nations today would probably do so.
- It is the doctrine of our major potential enemies in the Communist world to attack without diplomatic warning, and they almost certainly would do so, however much generalized political warning there might be beforehand of their intentions. In the short term, this is one of the easiest and most common means of deception (see Chapter 29).
- The pressures brought by other states, through the United Nations or otherwise, to forestall conflicts are such that nations today increasingly

> feel compelled to act without diplomatic warnings so that the international peacemaking machinery will not be brought to bear before they achieve their objectives.

Altogether, it is probable that specific warning of impending attack through diplomatic channels is largely a thing of the past. This does not necessarily mean, however, that more generalized indications of intention will not continue to be evident through the conduct of foreign policy and diplomacy. Indeed, as the dangers and costs of war increase, there is considerable reason to believe that there will be ample evidence that international political relations are seriously deteriorating before wars break out. In other words, we should still expect generalized strategic warning, if not short-term warning of imminent attack, from such developments. This may, however, require increased sophistication of analysis to recognize that war is imminent.

Propaganda Analysis

The term "propaganda" here is used in its broadest sense, to cover all information put forth by any means under national control or direction, which is designed to influence the intended audience. Propaganda can be either true or false, or somewhere in between, and it can be intended for domestic or foreign consumption or both. It can be disseminated through private channels (i.e., to the party faithful or cadre in briefings, directives, or "resolutions") or, through the mass media to the domestic population or the world at large.

The potential value, and difficulties, of propaganda analysis for assessment of intentions (that is, for warning) are well recognized. Propaganda analysis became recognized as an art, if not a science, during World War II, when specific efforts were made to analyze Nazi pronouncements for indications of possible forthcoming German military moves, as well as for other purposes. This apparently was only a partial success, in that propaganda proved not to be a very specific guide as to what the German Army might do next, or when, although it did provide some useful insight into how the Germans viewed the war in general—and thus was of value in judging whether a new offensive effort might be brewing.

Propaganda analysis in the US government today is diffused throughout the community, and in fact most intelligence personnel probably feel themselves qualified to some degree to interpret what our adversaries are saying, which may be one of our problems. For there are in the intelligence system some offices and personnel with specific qualifications and experience in this field, whose views have not always been given as much attention as they deserve. Certain overseas diplomatic posts—most notably Moscow for the USSR and Hong Kong for China—have concentrated on this type of analysis and have generally excellent records in interpreting the significance of propaganda pronouncements of our two major adversaries.

On the home front, the Foreign Broadcast Information Service (FBIS) has a propaganda analysis staff which is devoted almost entirely to examining and reporting trends and new developments in Communist propaganda and which concentrates on "identifying new elements or departures from the norm, defining the toughness or softness of public statements and propaganda themes, isolating indications of policy shifts, sensitivities, or projected actions." In addition to continuing analyses of trends in general, major propaganda statements (such as Chinese Foreign Ministry statements, or authorized TASS statements) are analyzed by FBIS in depth, with careful attention to the significance of particular phrases as well as the general thrust. In addition, FBIS maintains what are known as War Themes files. From a warning standpoint, this title may be slightly misleading, since these are not primarily a compilation of phraseology or statements which have preceded acts of aggression or other crises but rather a collection of statements and data on attitudes toward war. Nonetheless, there is a wealth of material at FBIS to support the warning effort, both on a current basis and in depth.

The Value of Propaganda for Warning: A General Commentary

There are two prevalent misunderstandings about propaganda and its relationship to, and value for, warning.

First is the widespread tendency to mistrust or reject almost anything which our adversaries say as "mere propaganda" and hence to regard it as meaningless if not completely false. This tendency is particularly prevalent in the military services—a tendency which may derive both from their concentration on the military hardware and from lack of experience with the subtleties of political and propaganda analysis. This tendency to disparage the usefulness of propaganda is most unfortunate, for the record shows that propaganda trends, and specific pronouncements, are often very valuable indications of enemy intentions.

A second tendency, almost the opposite of the above, is to expect too much warning from propaganda, that is, to expect it to be highly specific or to provide virtually unequivocal evidence that military action is impending, perhaps even specific warning of the time and place. People who expect this kind of warning from propaganda are almost certain to be disappointed, and they may therefore conclude that the propaganda provided "no warning" when in fact the enemy's propaganda provided considerable indirect or less specific evidence of what he might do.

This writer is not an expert in propaganda analysis. The following general comments on the usefulness of propaganda are, however, derived from experience in many crises. We are here discussing the propaganda put out by closed societies through controlled media, where both its quantity and content are carefully regulated and designed to achieve specific goals:

Propaganda reflects concern. Propaganda is a very useful barometer of how concerned the nation's leadership is about particular issues. Marked upsurges in propaganda on a particular subject or area do generally reflect genuine preoccupation with it, particularly if sustained over any period of time. Similarly, a very low level of propaganda attention to an issue usually indicates very little concern with it. There are occasional exceptions to this under unusual circumstances. One might be when a particularly secretive issue was involved, such as delicate international negotiations. A second is when propaganda is "marking time" pending a decision by the national leadership on what to say, or do, about it; these lulls or drops in meaningful comment can signify that the issue is so important that all comment is being withheld pending guidance from the top. Finally, deliberate inattention to area or an issue can be used for deception, generally in the relatively short term (see Chapter 29).

Most propaganda is "true." We are here using "truth" in a relative, not absolute, sense. We mean that nations cannot continually distort their objectives and policies, and particularly not to their own people. A major function of propaganda in Communist states is to indoctrinate the populace, including the party cadre, in what they are supposed to think and what they are supposed to do. To put out totally false statements or misleading guidance is self-defeating and will not evoke the desired response. It may be observed that this is particularly true of great national programs or objectives, when a greater than normal endeavor or enthusiasm is being sought. It is important, when hostilities may be impending, to instill the proper degree of hatred or fear of the enemy, to persuade the people to work longer hours, to justify cutbacks in consumer goods, to encourage enlistments in the armed forces, and so forth. The leadership cannot afford to give a wholly false picture of the situation to the populace.

To illustrate the point further, there was a major argument in 1965–66 over the meaning and significance of a heavy barrage of North Vietnamese statements aimed at their own populace which called for large-scale enlistments in the armed forces, longer working hours, greater sacrifices, recruitment of more women so that men by the thousands could be sent "to the front" (i.e., South Vietnam), and so forth. There was a group in the US intelligence community which rejected all this as "mere propaganda" for our benefit and which would not credit it as evidence that North Vietnam was preparing to send large numbers of troops to South Vietnam. The contrary argument—which of course proved to be the correct one—maintained that just the reverse was true, that this intensive internal indoctrination was the true barometer of Hanoi's intentions, and that the official propaganda line (that there were no North Vietnamese troops in South Vietnam) was the false one put out for our benefit. The refusal to believe this internal propaganda campaign possibly was the single greatest obstacle to the recognition that Hanoi was mobilizing for a major military effort in South Vietnam.

Official authorized statements are unusually significant. The Communist press operates under a set of prescribed rules which have proved very consistent over a period of years. Routine, day-to-day events are handled under established guidelines; more important developments call for articles by particular commentators (sometimes pseudonyms for top officials); major issues evoke authorized or official statements from the highest level. These latter statements are important not only for themselves, but because they set forth the "party line" for the rest of the propaganda machinery and thus will be carefully adhered to by the faithful. These statements always warrant the most careful study and analysis, and when they may bear on war or peace they are of particular significance for warning. This does not necessarily mean that they will be easy to interpret. Some of these statements are masterpieces of Communist dialectic which presumably (we are not certain of this) are understood by those initiated into what someone has called "the art form," but whose true significance often eludes the rest of us. Whether we can or cannot, in any given instance, comprehend the dialectic will sometimes depend in part on the amount of study and attention given some of the details. It became clear in retrospect that some of the fine points in one of these Communist classics—the 11 September 1962 TASS statement which unknown to us really ushered in the Cuba crisis—were not given sufficient attention in the community.

There are also certain time-honored phrases which have connoted varying degrees of concern in the past, and hence may be general guidelines to what may happen again. History tells us that phrases such as the following have been associated with preparations for military moves or even with a firm decision to intervene with military forces:

- "cannot stand idly by"
- "will never allow [a given state] to be removed from the Socialist social system"
- "regards its security as directly threatened"

The above are illustrative, not a comprehensive list of such phrases. Of course, like other propaganda, such phrases cannot be interpreted in isolation but must be assessed in the light of all other evidence.

Propaganda warning is usually indirect rather than specific. As a general rule, phrases such as the foregoing are about as specific propaganda warning as we are likely to receive of military action. It will be observed that it would be difficult to be much more precise without describing specific military preparations (generally a no-no in Communist states) or directly threatening to intervene with military force, which is also usually avoided. I cannot recall an instance since World War II in which a Communist nation has publicly stated that it would intervene, invade or attack with its regular forces, even when such action was imminent. The closest any Communist country has come to such a direct statement was China's open calls for "volunteers" for Korea in the fall of 1950. In most cases, Communist nations like to maintain the pretense that their forces were either "invited in" or aren't there at all. China has never acknowledged that anything other than "volunteer" units were sent to Korea; North Vietnam never acknowledged the presence of its forces in South Vietnam until the 1972 offensive, when this was tacitly although not explicitly admitted; and Soviet troops, of course, were "invited" to enter both Hungary in 1956 and Czechoslovakia in 1968.

The commitment to act is highly persuasive. In line with the above, Communist nations are not often given to bluff or idle threats that they are going to take aggressive actions when they do not intend to carry out. We must hasten to add that there are exceptions to this, such as the direct Soviet threats in the 1956 Suez crisis to wipe out the United Kingdom with missiles, and a number of similar statements made later by Khrushchev, including several quite threatening ones concerning Berlin. Nonetheless, it is generally true that these states like other nations do not wish to entail the stigma of failing to carry out their threats, or promises, particularly when a public commitment is involved.

Therefore, developments in propaganda which convey a high degree of commitment to *do* something or to achieve a particular end are particularly meaningful. The public warnings of the Warsaw Pact states in July 1968 that Czechoslovakia would never be permitted to pass to control of the imperialists were unusually significant without saying what means might be used to prevent this happening; these statements implied that any and all means would be used as necessary. An even better example of the importance of such commitments for warning preceded the Chinese intervention in Korea. After a period of hesitancy and uncertainty, there was an abrupt shift in the international Communist propaganda line during the first week of November 1950 to an all-out support of the North Korean cause, which now became identified as the responsibility of the world-wide Communist movement. So striking was the change in the propaganda line that the US warning committee concluded that the major new development that week in the Korean situation was the shift in Communist propaganda to open acceptance of responsibility for the fate of North Korea, and that the unreserved nature of the propaganda implied an intent to turn the tide in Korea through an unofficial war by the Chinese Communists. This judgment preceded the massive Chinese offensive by almost three weeks.

Political Warning through Third Parties

No discussion of the type of developments which can give us political warning would be complete without some attention to the usefulness of intermediaries or third parties. This applies both when they are deliberately used as a channel to convey a message and when they serve as leaks, inadvertently or otherwise.

When deliberately used, it is often for the purpose of arranging discussions or negotiations, but it may also be to convey a direct warning. The Indian Ambassador in Peking was selected as the first channel to convey warning to the US that Chinese forces would intervene in Korea if US/UN forces crossed the 38th Parallel. Most people were inclined to dismiss this as bluff at the time.

Still more useful may be the unintended, or at least only semi-intended, leak through third parties. It is axiomatic that the more people, and particularly the more nations, brought in on a plan, the more difficult it is to keep it secret. Because of the very tight security on the introduction of Soviet missiles into Cuba, the USSR is believed to have informed very few foreign Communist leaders, probably only the heads of the Warsaw Pact countries under rigid security admonitions. This adventure was solely a Soviet show. On the other hand, the preparations for the invasion of Czechoslovakia required a high degree of cooperation and planning among five nations, and numerous people were cognizant of the general nature of the plans if not the details. In addition, the USSR elected some four weeks before it finally invaded to forewarn the non-ruling Communist Parties, probably throughout the world, of its general intention—i.e., to keep Czechoslovakia within the fold by any means required, including force. The result was that we learned far more about the Soviet decision-making process and plans than during the Cuban crisis (see discussion in Chapter 27).

Despite the rigid internal security of Communist states, their growing contacts with the West and the breakdown of the once monolithic system have served to improve our knowledge of what goes on within them. At the same time, many third world countries (such as the Arab states and India) have been courted by Moscow. China, as of this writing, is beginning to open up to foreigners. Over a period of time, our chances of obtaining warning information through various third parties will probably further improve. Although this is grounds for some optimism, the Communist nations almost certainly will remain capable of extraordinary secrecy, surprise and deception when they consider it essential to their security interests.

Internal Factors Assessing the Views of the Leadership

Few subjects have proved more elusive to us than a true understanding of the character, attitudes and proclivities of the leaders of foreign nations. This can be true even in countries with which we have friendly relations and numerous cultural contacts (e.g., France), When the nations are essentially hostile, or at least not friendly, and their leaders have been educated in entirely different traditions or ideologies, the potential for misunderstanding them rises dramatically. Of the Communist leaders since World War II, Khrushchev almost certainly was the most outgoing, garrulous and willing to meet with foreigners. If we thereby felt we understood him, we were disabused by the Cuban missile episode. Brezhnev, on the other hand, had apparently never met with an American other than Gus Hall prior to preparations for his Summit meeting with President Nixon in May 1972. Where the leadership is essentially collective (as it has been in several Communist nations), we often have very little if any perception of the lineup on particular issues. Despite several reports on the subject, we really do not know how the Soviet leadership voted with respect to an invasion of Czechoslovakia. Nor do we know which of the leaders of North Vietnam at any time have favored various tactics—e.g., prolonged guerrilla warfare versus large-scale conventional, offensive operations. It is possible to get widely varying opinions from professed experts on where General Giap really has stood on this question at various times.

As a general rule, therefore, the attempt to second guess our potential enemies based on any professed insight into their characters or attitudes is likely to be a risky business. When sudden changes in leadership occur and relative unknowns move into positions of power, the difficulties are compounded. The death of Stalin brought about a period of great uncertainty, and there was a brief period of intelligence alert against the contingency that the new Soviet leaders might undertake some hostile act. Of course, the reverse proved to be true. Stalin's successors were somewhat less hostile and aggressive and probably were partly instrumental in bringing about the armistice in Korea shortly later.

Nonetheless, there are instances in which a change in leadership has been of warning significance, and has been indicative of a change toward a more aggressive policy or even of a clear intent to initiate hostilities. The change of government in Japan in October 1941 is universally recognized as one of the key developments which foreshadowed a more aggressive Japanese policy. The formation of a new government under the militarist General Tojo set in motion the chain of events which culminated in the attack on Pearl Harbor. A somewhat less dramatic change in Israel five days before the Six-Day War increased the likelihood that Israel would initiate an attack on the Arab states. On 1 June, General Moshe Dayan, a recognized hawk, became Minister of Defense. US indications analysis identify the appointment of individuals known to favor war as a development which may foreshadow hostilities.

Coups and Other Political Surprises

Intelligence personnel, particularly the chiefs of intelligence agencies, become used to being blamed for things for which they were not responsible and which they could not conceivably have predicted. (This is partly compensated by the mistakes made by intelligence which are not recognized or brought to light.)

Nothing is more exasperating to members of the intelligence profession than to be charged with failing to predict coups and assassinations, which they rightly consider as "acts of God" somewhat less predictable than tornadoes, avalanches and plane hijackings. It is ridiculous and grossly unfair to expect the intelligence system to anticipate such acts, which are plotted in secrecy and sometimes by only one individual. Indeed, the likelihood that intelligence will learn in advance of a coup or assassination attempt is in inverse ratio of probability to the likelihood that it will come off, since the leak which reaches the intelligence agent will probably also reach the intended victim who will in turn take steps to forestall the attempt. Some veterans in the business are still irritated by the recollection of the investigation of intelligence reporting which followed the assassination of Jorge Gaitan in Bogota in February 1948, an unforeseen event which precipitated disastrous riots and unfortunately coincided with a Pan-American conference being attended by the US Secretary of State.

Forecasts of this type not only are not within the province of strategic warning, however serious the consequences of such acts may be. They are really not within the province of intelligence at all. The most that can reasonably be expected is that the intelligence system recognizes that, in certain countries or situations, such acts if they should occur might precipitate riots, revolts or other crises inimical to our interests. But even this is expecting a good deal, as the police record in this country of attempting to anticipate urban riots should demonstrate.

22

Economic Indicators

This chapter is concerned with those aspects of war preparedness which affect the civilian economy and civilian populace, as opposed to military logistic preparations and mobilization, which were addressed in separate chapters in Part IV. In practice, of course, this distinction is somewhat artificial, since in wartime the military and civilian preparedness measures are closely integrated, in fact are opposite sides of the same coin. Moreover, our ability to recognize that certain military mobilization and logistic preparations are under way is often largely a result of the impact of these steps on the civilian economy, so that from the analytic standpoint the military and civilian developments must be considered as a whole.

Economic indicators are potentially a very important aspect of warning but nonetheless are usually given relatively little attention. Most indicator lists carry comparatively few economic items in comparison with the voluminous numbers of military items, and there is some tendency to regard economic developments as either too long-term or too unspecific to be of particular pertinence for warning.

This attitude probably is the result in part of the types of wars—and the types of warning problems—which we have had since World War II. The great majority of the crises in this period have been either of a short-term nature (sometimes only a few weeks) or they have been long simmering cold war problems (such as Berlin) which did not result in hostilities. There have been only two important extended conflicts—Korea and Vietnam/Laos—which might have required a major diversion or recommitment of economic resources, and the Asian nations involved in these conflicts have relatively primitive economies in which the conversion to a wartime economy was both less drastic, and less evident than in highly industrialized states. Both Korea and North Vietnam, moreover, were heavily dependent on their major Communist allies for the supply of much of the ordnance and other materiel needed to sustain the war, so that many of the classic economic indicators were not applicable. Finally, both these wars escalated somewhat unexpectedly to major and extended conflicts; neither North Korea nor North Vietnam initially believed that they would have to expend so much in manpower and resources. Particularly in the case of Vietnam, there was a rather gradual upping of the ante brought about in large degree by the change in the character of the war itself.

In short, there has not really been an instance since World War II which illustrates the type and scope of economic preparedness measures which might be undertaken by a highly industrialized state in expectation of major hostilities.

Long-Term Fundamental Economic Indicators

The importance of major economic indicators derives primarily from the fact that they reflect the basic allocation of resources of the nation, toward war or toward peace. They are a tangible measure of national priorities, of the direction in which the country is headed. The shift from a peacetime to a wartime economy, particularly in industrialized societies, obviously entails an enormous and extremely expensive reallocation of national resources. Such basic changes in allocations and priorities will usually occur only if the nation's leaders are themselves preparing to initiate the conflict (e.g., Hitler's conversion of Germany to a war economy over a period of more than six years), or if they become convinced that there is a grave danger that they will be victims of aggression. In the past, nations often have been extremely reluctant to convert to wartime production or to initiate other major economic reallocations and have deferred the undertaking of such steps until war was already under way. This, of course, can reduce their military capability to respond and may prolong the conflict (e.g., the years required in World War II to build up sufficient Allied capability to take the offensive against Hitler).

One useful measure of the importance and validity of an indicator is how expensive it is to implement or accomplish it, how large a portion of the national resources is involved, or how much it inconveniences or reduces the standard of living of the citizenry. Clearly, by this standard, major economic reallocations which reduce the availability of consumer goods, increase taxes, entail longer working hours or otherwise have a major adverse effect on the pocketbooks and pleasures of the populace should be rated as highly significant indicators. And they are. It is safe to say that no nation undertakes steps of this type as a bluff, and that a determination to carry through with such measures is a clear and reliable sign of a genuine belief that there is a growing danger of hostilities.

The threat of a devastating but short nuclear war has raised questions whether some of the traditional long-term economic indicators should still be considered. applicable, particularly those which involve a major conversion of industry, production of large quantities of additional weapons, and so forth. The point of course is well taken, insofar as one is describing actions which would be undertaken either after the war had started, or not until it appeared imminent. From an indications standpoint, however, the possibility of nuclear war may raise the importance of such developments—in that the potential aggressor might decide to undertake such measures well prior to the start of the conflict, rather than delaying them until it might be impossible to carry them out. There seems little doubt that the nature of modern war has increased the importance of initial preparedness, both offensive and defensive, and therefore could increase the likelihood that the aggressor will feel compelled to take more steps prior to hostilities (and thereby hopefully give us more indications) than was the case in the past. One such key preparedness measure for modern war—civil defense—will be discussed in the next chapter.

Short-Term Economic Indicators

Apart from a major real location of economic resources, there are a variety of less drastic economic measures which may be undertaken in preparation for, or in expectation of, hostilities. Developments of this type have sometimes proved extremely valuable indications of impending hostilities. Two examples from the period since World War II will illustrate the point.

In November 1950, as Chinese Communist troops were preparing for their offensive in Korea, a series of reports indicated considerable confusion and changes of plans in Communist commercial circles in Hong Kong. Purchasing agents apparently had been instructed by Peking to concentrate on items readily available in Hong Kong and to withhold further purchases from overseas. Chinese companies sought to cancel contracts with delivery dates later than 30 November, and they inserted clauses in contracts providing that the buyers could change the port of delivery if hostilities were in progress. In several cases, funds were transferred from Hong Kong and the US to Swiss banks, and efforts were made to finance Chinese purchases through Switzerland. On at least two occasions, Chinese Communist officials indicated that if a full-scale war did not occur in December, the immediate threat would be over and buying might be resumed. In sum, the evidence strongly indicated that Peking expected that war in the near future would result in the elimination of Hong Kong as a trading center. These developments, which were most unusual and have never since occurred, were among the strong indications that Peking was preparing for an early and major military offensive in Korea.

The second illustration—the mobilization of Soviet reserve transportation units prior to the invasion of Czechoslovakia—has been discussed elsewhere in this handbook (see Chapter 16). What made this effort unusually significant as an indication of Soviet intent was the season of the year, for normally during the harvest season this pattern would have been reversed—i.e., military personnel and trucks would have been assigned to assist in the harvest. Thus, the development indicated an overriding military requirement for truck transport even at the cost of possible failure to bring in all the harvest on schedule.

Examples of Key Economic Indicators

From the multiplicity of possible economic developments, there are certain types of things which the indications system attempts to watch and which have been judged to be of potential value for warning. Some developments naturally will have greater specific warning value than others, and therefore some assessment of their potential usefulness and of our ability to collect the information on a timely basis is also included in the following discussion. Since we have had so little experience with some of these indicators since World War II (particularly in the Soviet Union which has not been mobilized for war), it must be understood that forecasts in this field are perhaps even more hazardous than in some other aspects of the warning business.

National budgets and defense allocations. These are long-term basic indices of national economic priorities which can give us a sense of what proportion of the GNP is being allocated to military matters and whether the trend is up or down. A marked change involving greatly increased appropriations for weaponry is likely to be discernible even in countries, such as the Soviet Union and its Eastern European allies, which conceal many of the details of the budget. Careful study by trained economic analysts also frequently provides other useful data on trends in allocations for specific military or military-related matters—such as the total manpower in the armed forces or appropriations for research and development. The basic data and the accompanying analyses thus can be very useful barometers of general trends toward a wartime or peacetime economy, and even of more specific use in pinpointing particular economic allocations. Needless to say, this may be invaluable to us over the long term. Presumably, a full conversion to a wartime economy might be reflected in published budget figures. Some Communist states, however, have revealed so little about the national budget that it is of little indications value, and any closed society may choose to withhold vital data (or possibly even falsify them, although this is less likely) if it considers it to be in the interests of national security.

At best, such developments are not likely to provide us much specific warning that hostilities are likely in any given time frame, or of where they may break out. There has been a tendency in recent years for the warning system to pay very little attention to this type of development, although formerly the Soviet budget was annually examined for any trends or items deemed to be of military indications significance. One reason perhaps is that so much of the Soviet military budget in recent years has been allocated to the long-term buildup of strategic capabilities, largely handled in detailed national estimates. Were Soviet military or military-related allocations to increase or otherwise change substantially at some time in the future, it is likely that more attention would again be paid to this type of development from a warning standpoint.

Shifts in economic production. These are closely akin to the preceding item, and in fact a reflection of the allocations in the budget. Thus, they have much the same value as long-term indicators of military trends. Changes in economic production of the "butter to guns" type, however, may provide more specific information than the budget and can be useful indications of the nature of military preparations for possible impending hostilities. Substantial increases in the production of tanks, aircraft, small arms and ammunition, missiles, submarines and the like are not only measures of an increasing military capability; they may also be indications that a nation is preparing to initiate war at some time in the future. Sudden and urgent changes may be even more useful as indications: the initiation of round-the-clock production in military plants, sharply increased employment of women in jobs usually held by men, or marked increases in production of specific items which might be critical for the military in the event of war but of little need otherwise.

Experience has taught us that we do not learn much about what goes on inside major production facilities of closed societies until we obtain evidence that they are turning out particular items of equipment. We are not likely to learn that an order has been issued to convert them from production of farm tractors to armored personnel carriers, and it will probably be quite some time after the conversion has been accomplished before we will learn that APCs are being produced there. In some cases it has been a very long time. In general, the larger and readily hidden the item is, the sooner we will learn about it. We are much more likely to pinpoint the aircraft factories and have some estimate of their monthly production than of the ball bearing factories.

The Soviet Union, and to a lesser extent other Communist states but particularly those in Eastern Europe, are known to have detailed industrial mobilization plans. These include military mobilization departments in the various national ministries and specific mobilization plans for industries and individual plants. Obviously, evidence that any plant or group of plants was implementing its mobilization plan would be of critical importance for warning. It is very difficult to forecast what our chances would be of obtaining such information on a timely basis, but experience suggests that there might be considerable delay in the receipt of such information from within the USSR, at least for specific individual plants. In Eastern Europe, the prospects for obtaining such information are considerably better.

Stockpiling and strategic reserves. As a corollary of their mobilization systems, the Communist states (again, particularly the Soviet Union and its Warsaw Pact allies) have placed great emphasis on the stockpiling of large numbers of critical items which might be needed in the event of war. In the USSR, these stockpiles are known as State Reserves. Although we know little about the size of the reserves or how the system has operated in recent years, there is every reason to believe that the USSR and its allies have placed continuing and probably increased emphasis on the building up of strategic stockpiles for wartime use. Our evidence suggests that their concern with the "initial period of war"—which they view as critical if not decisive to victory—has resulted in the most careful buildup of stocks of innumerable items which would be available both for regional use and for supply of the military forces and specific plants or installations. Presumably the existence of such stockpiles will reduce the need for urgent accumulation of goods in the period preceding the outbreak of hostilities, including that interval which the Communist nations describe as the "period of threat." Nonetheless, no system operates perfectly, and it is logical to expect that there would be some intensive effort to improve and replenish reserves, and perhaps to relocate some of them, prior to the outbreak of major hostilities. It is also possible that some reserves would be released in advance of hostilities to fulfill some urgent requirements. It was learned after the outbreak of the Korean war that the USSR had released some materiel from State Reserves to support the initial stage of that conflict. On rare occasions, some items from Soviet strategic stockpiles also have been released following natural disasters. Any seeming abnormal emphasis on the buildup of strategic reserves or evidence of unexplained releases thus might be an indication of impending hostilities.

The stockpiling of strategic items also may be reflected in changes in imports and exports, particularly urgent efforts to acquire strategic materials from abroad and the cessation of exports of such items. Obviously, our chances of obtaining such information are much better than on internal developments. Although knowledge that a nation is making intensive efforts to acquire some specific item or items is not, of course, necessarily evidence of hostile intentions, it can in some cases contribute substantially to such a judgment. For example, heavy North Vietnamese purchases of drugs and other medical supplies in the 1965–67 period was one of the major indications of preparations for the large-scale commitment of forces in South Vietnam.

Transportation. This subject was discussed in Chapter 16, where it was noted that the Communist nations have extensive plans for the requisitioning or takeover of civilian transport in the event of war, and that disruption of normal transportation is likely to be one of our best indications that military forces are being mobilized and deployed. Moreover, it is a development which we have a good chance of detecting, particularly in the forward areas such as Eastern Europe, but also in some degree in more remote areas. Most of the Communist world remains heavily dependent on rail transport (far more than the US), so that traditionally our best indications of abnormal military requirements on the transportation system have been reflected in rail traffic. As trucks have come into greater use for military movements, the requisitioning of trucks from the civilian economy (as in the invasion of Czechoslovakia) has proved as useful an indication as disruptions of rail traffic. The steady rise in air transport has substantially increased the likelihood that the takeover of civilian aircraft for military use will be a significant indication of approaching hostilities, and that we shall have some evidence of this. Altogether, transportation has been and promises to remain one of our best indicators of hostilities—and the more sophisticated and mechanized the military forces, the greater will be their needs and the greater the impact is likely to be on the civilian population.

To ensure the rapid restoration of damaged transportation lines, the strategic reserves of both the national economy and the military forces almost certainly will include stocks of rail ties, bridging equipment and other items. The prepositioning of such reserves at bottlenecks in the transportation system would be another significant, and potentially observable, indication.

Food and Agriculture. Time permitting—that is, where preparations for war last for several months or more—there may be some changes in agriculture. These might include various efforts to intensify production and increase yields, requisitioning of a greater portion of the crop for national stockpiles, and the greater encouragement of private gardens and self-sufficiency by the populace. A recent article in the Soviet press entitled "The Problem of the Survival of the Economy in Modem War" emphasized that the survival of the economy in such a war would be "inconceivable without stable agricultural production."[1] Among the points emphasized was the need to protect animals from weapons of mass destruction, to include the preparation ahead of time of special shelters.

The likelihood that we would receive indications of this type—and correctly interpret them—is probably rather low, and moreover they would probably occur only in the event of pre-planned nuclear hostilities. A more meaningful indication—which is both likely to occur in preparation for more limited hostilities and which we may detect—is a shortage of food (particularly choicer items such as meat) for the civilian populace as such items are taken over in increasing quantities for the military. Actual food rationing is a measure which most nations are reluctant to impose even when significant shortages exist, and it might be that rationing would not be introduced prior to the outbreak of hostilities.

New economic controls and bottlenecks. A changeover toward a wartime economy is almost certain to require the imposition of some new economic measures or controls which will likely have an early impact on the average civilian. It will also produce shortages of items, whether real or artificial as the result of hoarding, of which the ordinary housewife becomes immediately aware—and which therefore cannot be concealed. Even minor international crises have produced waves of panic buying in the Communist countries of Europe. Shortages of some commodities and items have been so frequent that it may be difficult to tell whether some new military requirement has produced the shortage, or whether it is just another manifestation of a chronic problem. Although new economic regulations and bottlenecks are therefore potentially useful indications, it may not be possible to discern whether the indication is a real manifestation of a wartime measure. In the event of widespread and extensive economic preparedness measures, however, it would seem likely that their real cause would be apparent. Full economic mobilization, like full military mobilization, is a step which no nation could hope to conceal.

Note

1. This article appeared in *Communist of the Armed Forces*, February 1972, pp. 9–16.

23

Civil Defense

A SEPARATE CHAPTER ON CIVIL DEFENSE is warranted because this aspect of preparedness has become so important to modern war, particularly nuclear war. There is probably no other aspect of preparation for war—except possibly the changes in doctrine and tactics for the conduct of nuclear war—which has changed so extensively in the period since World War II. Moreover, the changes in the nature of civil defense preparations—at least in those countries with which we might conceivably become involved in nuclear war—are such that they have dramatically increased the involvement of the population as a whole, and consequently have greatly increased the likelihood that there would be meaningful indications in the civil defense field prior to the outbreak of hostilities. In fact, some students of this subject believe that civil defense is the most likely field of Soviet preparedness to provide us strategic warning, assuming of course that the USSR actually carries out the measures which it has planned to implement for the defense of its population. The extent to which it would or would not do so—and the arguments pro and con—are discussed later in this chapter.

There is a great deal of literature available on the Soviet civil defense program, and a lesser amount on the programs of the other Warsaw Pact countries and the Communist nations of the Far East. Much of the information is unclassified, since the Soviet press is the single best source of information on the program. Although some aspects of the Soviet program are secret, most of the information has been given extensive publicity inside the USSR. There are numerous intelligence studies of various classifications on the Soviet civil defense program, which the student of this subject may consult. Therefore, this discussion will attempt to outline only the highlights of the program, as background for some assessment of what we might expect to see in the event of actual preparations for a major conflict. We will also examine, much more briefly, the civil defense programs of other Communist states.

Highlights of the Soviet Civil Defense Program

The USSR has by far the most extensive, elaborate and well-organized civil defense program in the world today. In comparison with most of the West, where civil defense planning is still in its infancy and the merits of any program to protect the civil populace remain in dispute, the USSR has moved forward with a massive civil defense program. It is directed by an extensive staff which supervises an elaborate organization reaching throughout the governmental and economic system and which involves almost every citizen of the country. Following are some of the more important aspects of the program.

Organization. For more than a decade, the Soviet civil defense has been directed by a joint military/civil organization (whose chief has been a Marshal of the Soviet Union) directly responsible to the Soviet Council of Ministers. Major responsibility for the implementation of the program lies with the military which provides most of the personnel for the "civil defense staffs" assigned at all levels of government and throughout the economic system. The civil administrative chain of command is from Moscow down through the governmental (not party) structure to the lowest echelon. Each Soviet military district also has a deputy commander for civil defense and a staff which would assume direction of the civil defense program in event of war. Schools have been established at the national level for the training of civil defense officers, and civil defense units, both civilian and military, have been formed, trained and equipped. The scope of the program is illustrated by listing some of the functions which these units would perform: direction of nation-wide and local communications and alerting systems; supervision and control of evacuation; transport service for evacuees, for the injured and for essential supplies; medical services; firefighting; engineering services, including the construction of shelters; restoration of electric power and other utilities; maintenance of public order. The civil defense organizations in economic enterprises and agriculture have appropriate similar functions. The farm units have several special tasks, including protection of livestock, crops and water supplies, and the reception and accommodation of urban evacuees. These services, which require thousands of full-time personnel in peacetime, are expected to involve millions of people in time of war.

Training. Civil defense training is compulsory for almost everyone in the USSR. It begins in the elementary grades of the schools and extends upward through the educational system. It is mandatory in factories and agricultural enterprises, and unemployed adults are not neglected. A series of nation-wide civil defense training programs since 1955 probably have reached nearly all citizens. Increasing emphasis has been placed on the training of instructors and specialized personnel in recent years. Although the Soviet press has periodically complained of public apathy and inadequacies in the program, the quality and effectiveness of training is believed to have improved significantly in recent years. If it is possible to protect the civilian populace at all from nuclear war, there is little doubt that the USSR is the best prepared and organized of any major nation, probably of any nation in the world, to do so.

Shelters. At one time, the USSR placed considerable emphasis on the construction of underground shelters, and the basements of many apartments constructed during the 1950's were specifically designed as such shelters. This program was abandoned in 1961, after studies determined that such shelters would provide scant protection against more powerful and accurate nuclear weapons, and a program of mass urban evacuation was substituted. It is expected, however, that shelters would still be used for some personnel. Elaborate shelters almost certainly exist for top party and government personnel—quite likely both in and outside the Moscow area. Shelters also will be available for factory workers who are not scheduled for evacuation. The Moscow subway, long considered a likely bomb shelter, probably would be so used even if a major evacuation of Moscow were accomplished. In rural areas, there are almost certainly insufficient shelters to protect both the local populace and the urban evacuees, and a crash program to construct simple fallout shelters in rural areas is expected in event that a major conflict is threatened.

Evacuation and dispersal. The most ambitious, complicated and difficult aspect of the Soviet civil defense plan is the program to evacuate the majority of urban inhabitants to small towns and farms. Plans apparently call for the evacuation of about 70 percent of the residents of all cities over 100,000 population, with the remaining 30 percent to stay behind to man essential industries and services. Moreover, it is the Soviet expectation that these evacuations would occur for the most part prior to the outbreak of hostilities—an expectation which obviously would entail a period of strategic warning. Elaborate plans have been drawn up for the evacuations. These include: the selection of the modes and routes of transport, with the railroads to provide the bulk of the transport; selection of the dispersal areas and embarkation and debarkation points; preparations for the issuance of evacuation cards to those who are to depart the cities; instruction of individuals in what they are to take with them (a three-day supply of food, plus a limited number of personal belongings); the study and planning for numerous details which would be involved in the carrying out of such a massive resettlement program; and the conduct of exercises by the civil defense staffs and selected units to train them in the implementation of the program.

The Soviet civil defense chief has stated that urban evacuation could reduce casualties from 80 or 90 percent of a given city's population to less than 10 percent. Even if this is a somewhat exaggerated assessment of what the USSR actually expects (and does not, of course, take account of rural casualties or other factors), it appears clear that the Soviet leadership believes that the evacuation program could greatly reduce casualties from an initial nuclear strike and furthermore might be decisive in insuring the survivability of the economy, the country, and the socialist system.

The Soviets have not specified publicly just how much time they expect to require to evacuate a major city, although they have specified that about six hours of alert time would be needed before the first evacuees could depart. Some independent US studies have been made of the feasibility of evacuating a number of selected Soviet cities of more than 100,000. These have concluded that, under favorable conditions (which include no loss of available rail transport), 70 percent of the population of all of these particular cities could be evacuated in three to five days. Just how this might function if *all* Soviet cities over 100,000 were being evacuated simultaneously is, of course, another matter. Presumably, there might be considerable difficulty in areas of the western USSR, particularly in the general Moscow area, where there are large numbers of cities with more than 100,000 people. Nonetheless, these studies would indicate that the Soviet plans are not altogether impracticable, even though there would no doubt be numerous problems in actual implementation. From a warning standpoint, the most important point by far is that the USSR does really plan such a massive evacuation effort and moreover seemingly expects (or at least hopes) that it will have at least several days in which to carry out the program before war starts.

Civil Defense Programs in Eastern Europe

In other matters, the Eastern European members of the Warsaw Pact have generally followed Soviet direction and precepts, and to some extent, this is also true in matters of civil defense. Just as war plans are closely coordinated, we would expect civil defense plans and preparations to be similar and the actual implementation of such programs to be more or less simultaneous. The Eastern European countries have, in fact, given a fair amount of public attention to civil defense programs and the training of civil defense personnel. Nonetheless, the conclusion seems inescapable that these programs generally have lagged well behind that of the USSR and have involved much less ambitious plans for the protection of the civilian populace. Apart from psychological and budgetary factors, one major reason for this probably is the smaller size and greater population density of these countries, which would seem to make massive evacuation programs a considerably less practicable means of protecting the populace than in the USSR. A Czechoslovak colonel who was engaged in various aspects of Czechoslovak defense planning has described an extensive evacuation of the civilian population and economy as a practical impossibility and of very questionable effectiveness. This may reflect actual Czechoslovak official attitudes, and possibly those of other Warsaw Pact nations. Nonetheless, the civil defense plans of these countries apparently call for some dispersal from the cities, as well as hardened shelters for key personnel and other protective measures. Several Eastern European countries are known to have conducted occasional civil defense exercises involving some evacuation measures. A Hungarian announcement of an exercise conducted in June 1969 stated that over 80,000 persons would take part, including both civilian units and military formations, and that it would involve "tasks connected with evacuation, the provision of accommodation, and rescue, relief, and rudimentary restoration work."

How Would Civil Defense Actually Function?

The potential difference between theory or doctrine and actual practice is nowhere more evident than in the field of civil defense. Opinions as to what the USSR would *actually* do in the civil defense field, if major hostilities threatened, range all the way from those who believe that it would do almost nothing which would be obvious to us and certainly would not undertake massive evacuations of the cities, to those who maintain that a full implementation of civil defense plans is to be expected, time permitting. Many of those who hold to the "do-nothing" theory—on the grounds primarily that it would provide us such clear-cut warning that the USSR would not tip its hand—also maintain that the evacuation program would actually be impossible to implement and therefore really would not be attempted on any scale.

Obviously, there is no answer to this question which is going to satisfy everybody. It is the same problem as how much buildup there would be of Soviet forces in Europe prior to attack, how much political warning we would have, and so forth. How much would weight of attack be sacrificed to achieve surprise? Would protection of the civilian populace be more important to the Soviet leadership or would it be more important to them to avoid taking measures which we would almost certainly view as very ominous, perhaps would interpret as evidence that the USSR was planning a preemptive nuclear strike?

The Soviet performance thus far has made it evident that massive live civil defense preparations, including evacuation of cities, are regarded by the USSR as extremely serious steps which therefore probably would not be undertaken unless the leadership was convinced of a grave danger of hostilities. Our relative confidence on this score derives from several considerations: such a program is tremendously expensive and disruptive economically; it would be certain to have a most serious and adverse psychological impact on the population, the consequences of which might be unpredictable; the USSR could not be sure what the US/NATO response would be to an "exercise" of such massive and realistic proportions; and, so far as we know, the USSR has yet to undertake a massive evacuation of any city, either for exercise purposes or as a genuine precautionary measure during a crisis. It is of interest that, even during the Cuban missile crisis when Soviet air defense forces were brought to a very high degree of alert and Khrushchev was genuinely apprehensive that war could occur, there were no major civil defense preparations taken in the USSR. Undoubtedly, the civil defense staffs were alerted and certain other preparedness measures were probably taken relatively unobtrusively, but there was no mass alerting of the population.

This means, of course, that we also have no basis on which to judge how effectively the evacuation program might function, even if undertaken as an exercise in a period of calm. It is possible that the USSR some day may test out the program in a few selected cities, probably ones to which we would have little access. It is conceivable that it has already conducted a mass evacuation of some city, but we believe it unlikely that some news of this would not have trickled out in time. Such actions, however, probably could be effectively concealed in many cities of the USSR for at least a short time, provided the radio was not used to alert the populace. It is doubtful, however, that it could be done simultaneously in a number of cities without some knowledge of it reaching us. Obviously, it would be impossible to conceal a mass evacuation of Moscow, or probably of Leningrad.

In the event of a decision to conduct a mass evacuation because of a threatening international situation, it appears highly unlikely that it would not be announced to the Soviet public through radio, television and press. Reassurances from the leadership and repeated public instructions as to what to do would appear almost a necessity to prevent chaos, no matter how much advance planning and rehearsals by civil defense units there might have been. The maintenance of public morale could be critical to the military and economic effort, and indeed a prime objective of the civil defense program is to prevent a breakdown of public support so that the military effort can be sustained. Thus it appears most unlikely that some public explanations of the reasons for such drastic measures would not be forthcoming. It is probable, in fact, that the entire propaganda machinery would be called into action to explain or justify the civil defense effort.

In addition to uncertainty as to the mass psychological reaction, it appears highly questionable that the Soviet transportation system would function anywhere nearly as effectively in practice as theoretical studies of the evacuation problem would suggest. This is not necessarily because Soviet transport is inefficient (actually the railroads function quite well), but rather because there would be such heavy competing demands from the military for the transportation. There would obviously be all sorts of other difficulties in the implementation of an evacuation program, particularly the feeding of the populace in the countryside for any length of time, which would pose enormous problems and would almost certainly impair the smooth functioning of the operation. No doubt, however, the USSR would not expect it to function exactly as planned, and presumably would regard the program as a success if only half the evacuees ultimately were saved from nuclear destruction.

All this, however, does not answer the question whether or not the USSR would in fact attempt to carry out such a program. If there can be no demonstrably correct answer, there can be certain general judgments as to what would appear to be the most likely course of action. As in other aspects of the warning problem, there will undoubtedly be those who will not agree.

The most careful students of the Soviet civil defense program are generally persuaded that the USSR would implement the program in the event of a grave threat of hostilities with another nuclear power, i.e., the US or China. Moreover, it would appear likely that the program would be implemented both in event the USSR planned to strike first or feared that the enemy was preparing to strike. The arguments in support of this view are as follows.

The massive amount of effort which has been put into civil defense planning and preparation would not have been undertaken unless it was intended to implement the program in event of approaching hostilities. The subject must have been extensively debated already and the conclusion must have been reached that the program is both feasible and necessary to the Soviet war effort.

The necessity for the program does not derive solely or perhaps even primarily, from concern with saving the lives of more people. It derives from a conviction that the military effort will be dependent on the maintenance of a viable civilian economy, and that the Soviet Union could not survive a nuclear conflict in which masses of the urban populace were eradicated in the first hours or days of the conflict. Although Soviet leaders have wavered from time to time on whether any nation could survive a nuclear conflict and whether there could be any "victors," their planning indicates that they consider that their comprehensive civil defense program will provide them a better chance for survival than would otherwise be the case.

The civil defense program is an integral part of the war plan. The military establishment is deeply involved in it, and the militarized units of civil defense have prescribed roles to perform under the war mobilization plan. Thus, to omit the civil defense portion of the plan would require the reversal of many standing military orders and the reassignment of many reservists to other functions. It would further require the issuance of entirely new instructions to innumerable party and government officials and plant managers who now have prescribed functions to perform in the civil defense program and who have rehearsed what they are to do. In short, a decision not to implement the civil defense program could require a revision, at the last moment, of the war plan. As one writer on the subject has put it:

> It appears that a decision to omit civil defense would be administratively as complex as a decision to cancel participation of aircraft in an air defense effort and leave the job entirely to missiles. . . . With so many people involved, the planners of the strike have a problem: would the security of the surprise be well served by an attempt to leave out civil defense?
>
> Most important is the probability that the party leaders would not accept a military plan which excluded civil defense participation. One totally unacceptable result of such a plan might be the decimation or worse of the party while the military leadership remained relatively unimpaired. Another consideration of the Presidium ought to be the reaction of the surviving members of the populace, as well as of the party, if available civil defense facilities had not been put to use.

Civil Defense Programs in Asian Communist Nations

In comparison with the USSR, the Asian Communist countries have relatively unsophisticated civil defense programs, although they have not neglected the problem.

North Vietnam, which has been engaged in war more or less continuously since World War II, undertook a civil defense program only after the start of US bombing in 1965. Fears that Hanoi, and to a lesser extent other cities, might be bombed led to fairly extensive efforts to evacuate non-essential personnel. These efforts proved rather ineffective, however, since the populace showed a tendency to drift back to the cities, particularly when it appeared that the danger to civilians was not very great. The resumption of US bombing in the spring of 1972 resulted in another partial evacuation of Hanoi, which also appears to have been a rather half-hearted effort. The North Vietnamese civil defense program also has included extensive use of improvised bomb shelters, many of them concrete pipes. North Vietnam's civil defense measures were of relatively little value to us as warning indicators, since a stepped-up effort usually followed the initiation of US bombing, rather than anticipating it.

In Communist China, civil defense efforts have been sporadic and usually can be directly associated with some immediate developments which have raised Peking's apprehensions that it might be subjected to attack. In the few weeks prior to the major intervention of Chinese forces in Korea in the fall of 1950, there was a rash of reports concerning preparations for evacuation of government offices or other facilities from Chinese cities and some increase in air raid precautions. So far as is known, however, few if any evacuations were actually carried out, and Chinese fears apparently abated when the US did not respond to the Chinese offensive in Korea by attacking Chinese territory. There is little indication that the recurring crises in the Taiwan Strait area or the brief Sino-Indian conflict of 1962 were accompanied by any serious civil defense planning.

The conflict in Southeast Asia—particularly the Tonkin Gulf incident in August 1964, the initiation of US bombing of North Vietnam in early 1965, and the introduction of Chinese engineer and antiaircraft units into North Vietnam shortly later—raised Chinese fears of a possible US attack and led to the first of Peking's so-called "war preparations" campaigns. This ominous sounding phrase actually involved a variety of measures to persuade the populace to greater efforts and to instill support for the regime's programs—not all of them directly related to increasing war preparedness. Among the real preparedness measures was a considerable increase in civil defense activity, concentrated not unnaturally in South China, although the program also extended to other areas of the country. These preparations included: a considerable propaganda campaign to alert the populace to the danger of war "at an early date, on a large scale, with nuclear or other weapons"; the digging of air raid shelters, particularly trenches; numerous reports of plans to evacuate personnel and government offices from cities, together with a limited amount of actual evacuation; instruction in civil defense measures; air raid drills; stocking of food supplies; and a step-up in militia training.

This effort abated after two or three years, but the "war preparations" campaign was revived in 1969 following the clashes on the Sino-Soviet border. This time the danger of war with the Soviet Union occasioned a step-up in civil defense preparations, including renewed preparations to evacuate personnel from the cities and construction of air raid shelters. The program appears to have involved less intensive propaganda and participation of the populace than in 1965–66, but there were indications of more effective and permanent shelter construction. This, together with a program to decentralize industry into the interior, suggested that China was embarked on a serious long-term effort to improve its defensive capacity.

The history of the Chinese effort would suggest that increased emphasis on civil defense is a good indication of Peking's concern for national security from a defense standpoint. It is likely that civil defense preparations also would step up markedly in the event that China was planning for aggressive military action.

Little is known about civil defense in North Korea. There has, however, been an intensive effort for several years to put much military equipment into caves, tunnels and underground facilities. It would be surprising if some similar effort had not been devoted to shelters for civilians.

Conclusions

Our experience with civil defense indicators in several conflicts justifies a conclusion that the intensity of such preparations is generally an excellent barometer of whether a nation really believes it is in danger of attack. Large-scale and economically disruptive civil defense measures are unlikely to be initiated unless the threat is considered grave. Like other defensive preparations, a high level of civil defense activity may indicate either a fear of enemy attack or an intent to initiate the attack. The threat of nuclear war has greatly increased the importance of prior civil defense measures, particularly the removal of key officials to secure areas and at least a partial evacuation of civilians from urban areas. Nations which have extensive civil defense programs regard them as an integral part of their war plans, if not essential to survival, and it is therefore likely that these plans would be put into effect prior to the outbreak of a major conflict, if time permits.

24

Security, Counter-Intelligence and Agent Preparations

In this final chapter on indications in the civilian area, we shall deal briefly with the value for warning of developments in the field of security and clandestine operations. Although usually less exotic in actuality than in the world of fiction, the cloak-and-dagger business nonetheless has the potential of providing us with considerable insight into the enemy's preparations for hostilities.

Security Measures

There is no question of the importance of security measures in Communist nations for purposes of concealment. It is the primary means by which we are denied information about what is going on in much of their society all the time. The need for security dominates virtually every aspect of life and conceals from us a vast amount of basic social, economic and military activity. The two major methods by which this continuous blackout of information is accomplished are the control of the press and all other public media, and intensive physical security around most military and economic installations and activities, which in practice often means that large areas of the country are off-limits to travel by potentially unfriendly foreigners, both official and unofficial.

As a general rule, virtually all military installations (barracks areas, compounds, training areas, military headquarters, missile and antiaircraft sites, airfields, naval installations, depots) in Communist nations are closed to the public, not only to foreigners but to the local citizenry. Moreover, they are not just closed but intensively secured and guarded against unlawful entry. Equally drastic security measures surround important economic enterprises which are engaged in the production of military or military-related items. The rare occasions on which foreigners, such as attachés, are permitted visits to military units or to such factories are carefully controlled and usually are confined to selected show pieces.

In these nations, the "need to know" principle is really adhered to and indeed carried to fantastic lengths by our standards. A rigid compartmentalization of knowledge—within the military establishment, and in such fields as weapons production and research and development—denies details of the overall effort to all but a select few at the top. There is little doubt that these stringent measures impair efficiency and inhibit the free exchange of productive ideas in some degree.

The intended—and, to some extent, actual—effect of these measures is to deny us the basic data from which we can determine what is normal. If the system worked with total effectiveness, we would have little or no specific data about the strength and locations of military units, their weapons and capabilities, maneuvers and other training activities, except what was released to the press or otherwise made available officially. At the same time, however, most of these nations do not really want us to be totally ignorant of their military capabilities nor to close off foreign travel entirely, so in practice some limited access to their less important "secrets" is condoned most of the time. In actuality, there is considerable variation in the extent of territory which is normally closed to foreigners in the different Communist countries. While vast areas of the Soviet Union are permanently closed and periods of so-called "detente" have had little effect on this policy, some Eastern European nations impose relatively little restriction on foreign travel except in or immediately around important military and industrial installations. The most denied area in the world to the West has been North Korea.

The extent of permanent security restriction and past practice in crisis situations are two factors which could affect how much additional restriction might be imposed to deny us information during a period of unusual military activity, including preparations for hostilities. This is a subject on which we have had a great deal of experience, and it is therefore fairly safe to make some generalizations.

It is Soviet practice always to deny travel by official foreigners, and often by other foreigners as well, to an area in which unusual military activity is under way, specifically troop deployments and maneuvers. It is most unusual for any Westerner to observe any significant troop deployment, and Western military attaches are virtually always denied access to areas of maneuvers or other military movements. On occasion, the USSR has imposed very widespread travel bans. In June 1969, the Trans-Siberian Railroad was closed to nearly all foreigners as well as to official travelers. One traveler who did make the trip, perhaps because of some Soviet oversight, reported very heavy eastward military movement, and we ascertained nearly three years later that there had been a mobilization exercise along the Chinese border during this period. Another very widespread travel ban had been imposed in the western USSR in June 1968 for reasons which have never been established, but it has been suspected that there might have been some type of mobilization test preparatory to the subsequent invasion of Czechoslovakia. During the periods of known troop deployments for that invasion, the USSR repeatedly denied travel requests of Western attachés[1] although there were a few occasions on which they were able to observe a limited amount of troop movement. During the week of the Cuba crisis, the USSR also prohibited most official travel, possibly to cover a general alerting of its forces, since we do not believe that any units actually were redeployed.

The USSR also restricts travel by military observers in East Germany by imposing both permanent and temporary restricted areas. A temporary restricted area is nearly always declared for any important exercise activity, and sometimes for other purposes. A sizable area along the Czechoslovak border was continually closed by series of "temporary" restrictions for more than three months prior to the invasion of Czechoslovakia.

Our experience in this field warrants a judgment that the USSR almost certainly would impose further restrictions, and probably quite drastic ones, to cover the redeployments or other preparations of its forces in the period prior to the initiation of war. Although such restrictions do serve to alert us to an abnormal military situation, the Soviets quite evidently consider that security is better served by the restrictions than by permitting some of the activity to be observed.

The Eastern European Communist states generally have much less restrictive travel policies than the USSR, and the chance of detecting any significant troop deployments is much better, although areas of actual maneuvers are usually off-limits. Various devices also are used to prevent travel of trained military observers at crucial times. Nonetheless, major and widespread travel bans in these countries would be quite unusual and would probably not be imposed except in extraordinary circumstances. It is interesting to note that Poland did not even impose such restrictions in July–August 1968 when major Soviet forces were deployed into the country for the invasion of Czechoslovakia, although quite severe travel restrictions were in effect in the USSR and East Germany, and to a lesser extent in Hungary, to screen the deployments.

In the event of preparation for hostilities against the West, there is a good prospect that still more drastic and unusual security measures would be imposed. One such measure could well be tightened censorship on foreign correspondents, and measures to deny them access to their usual sources. A major and highly significant security measure, begun by North Korea three months before the attack on South Korea, was the evacuation of civilians from the 38th Parallel. Obviously, any similar step along that or any other border could be a very important indication.

The limited access of Westerners, particularly US personnel, to the Asian Communist nations to date precludes much generalization about what their security policies might be if diplomatic relations were established and these areas opened up to more Western travel in general. Their histories since World War II, however, justify a conclusion that they are all highly security conscious and would almost certainly impose severe restrictions to prohibit observation or other disclosure of any military activity which they wished to conceal.

How much warning value are abnormal security measures? The chief value perhaps is to alert us to an unusual situation, probably involving some military activity, so that other collection resources can be brought into play. Security measures, no matter how drastic, are not in themselves evidence, let alone proof, that military deployments are under way. Some further collection is essential to establish what is going on. The more extensive and drastic the measures are, however, the greater reason we have to suspect that something covert and potentially hostile is under way, unless there is some plausible alternative explanation. It is likely that the period prior to the outbreak of hostilities would be marked by extraordinary security measures of a nature rarely observed in peacetime.

Counter-Intelligence Domestic Aspects

The role of counter-intelligence and security is to protect the state from its enemies both foreign and domestic. In police or quasi-police states, the domestic enemy is nearly as important as the foreign and sometimes more so, and a substantial amount of the counter-intelligence effort is devoted to watching and when necessary restraining potential dissidents. (This is also true, one may note, of many states which do not lie behind the Iron Curtain. For example, it is not unusual in many countries for known troublemakers to be rounded up in advance of the visit of a foreign head of state.)

In the USSR, the power of the KGB (Komitet Gosudarstvennoy Bezopasnosti—Committee of State Security) has been reduced since the days of Stalin when it served as an instrument of terror, but it remains a large, secretive and ubiquitous organization with authority to pry into the lives of all citizens and the personnel to accomplish it. Occasional press items that some well-known dissident has been sentenced to prison or a labor camp are a reminder that the power of the secret police still permeates the Soviet state, barely concealed most of the time and sometimes blatantly evident. Less clandestine but also a major force for the preservation of order are the internal security forces, a branch of the armed forces. The pattern is repeated in all Communist states, with some differences in organization and functions of the various security forces.

These forces have major roles to play in the event of war, including the roundup and incarceration of dissidents and others suspected of disloyalty, as well as a variety of other security functions. It is highly likely that the authority and personnel strength of these organizations would be increased in preparation for war and that the leadership would rely heavily on them to ensure their safety.[1] An increase in KGB activity, both obvious and covert, is to be expected. It would probably include tightened observation of and restrictions on foreign diplomats, attaches and newsmen, but more importantly widespread measures to ensure against an outbreak of dissidence and sabotage once war had begun. A large-scale roundup of potential troublemakers, perhaps very quietly and very shortly prior to the outbreak of major hostilities, would be likely.

Foreign Espionage and Counter-Intelligence

Even more meaningful than internal security and counter-intelligence measures are some of the changes which might occur in operations abroad prior to the initiation of hostilities. We are speaking of the vast underworld of espionage, subversion, and other clandestine and covert operations on which our potential enemies spend enormous sums of money and employ thousands of framed operators. For most US intelligence personnel—those who work on overt collection or analysis—this part of the intelligence process is only dimly and very inadequately perceived. It is "the other side of the house," and a side vastly more secretive and compartmented than almost any other phase of the intelligence process. Nothing is more laughable than the popular concept set forth by some writers of mystery stories that most intelligence work consists in espionage and counter-espionage and that we are all "spies." For the fact is that most of us in the intelligence business have nothing to do with spies, our own or those of other nations, and we are likely to know precious little about the whole subject.

This writer is one of that generally uninformed majority. Such observations as follow, therefore, are not based on any special insight into the operation of foreign espionage systems and should not be so interpreted. In fact, much of what will be said is from unclassified sources.

A variety of historical factors contribute to the USSR's obsession for security, distrust of foreigners and conviction that it is encircled by enemies which are seeking to destroy it. They include the tradition of political repression inherited from imperial Russia, the hardships of the revolution and the difficulties of establishing the Soviet state, the armed intervention by Western states which sought to overthrow the Bolsheviks, the invasion by Nazi Germany, and the post-wars suspicion of the West engendered by the years of the cold war and the limited contacts of the Soviet leaders with Westerners. Above all, of course, the suspicion of the West derives from the secretive and conspiratorial nature of the Communist system itself and its own dependence on force or the threat of force to remain in power. The mere existence of democracies is a threat to it.

The result of these attitudes is that the USSR, and to a lesser but nonetheless. important degree other Communist states, devote an enormous amount of effort and money to foreign espionage and subversion. Indeed, in many areas of the world—particularly the underdeveloped nations—such activities are probably the primary function of Soviet embassies and other official representation. Moreover, the centralized control of all foreign activities and the ability to maintain secrecy permit the KGB to place its officers anywhere within the official and unofficial establishment abroad, and in almost unlimited numbers—subject only to how many personnel the host nation is willing to accept. There is little question that the expansion of Soviet missions abroad in recent years is in large degree attributable to an increase in espionage and other covert activities. A celebrated case—the defection of a KGB officer in London in September 1971—provided dramatic evidence of the scope of such Soviet activities. As the result of his reports, the British expelled 105 members of the official Soviet establishment in London for espionage activities. They comprised nearly 20 percent of the total, and included nine of the Soviet Embassy's eleven counselors and five of its twelve first secretaries. The agents were established in every type of mission, including the trade delegation, the Moscow Narodny Bank and Aeroflot. The British action was reported attributable in part to the fact that the defector and some of his colleagues were engaged not in normally accepted espionage activities but in preparations for sabotage in the United Kingdom in the event of war. Numerous less dramatic examples of the all-pervasive nature of the USSR's foreign espionage program could be cited, and many of course have become common knowledge as the result of defections and arrests in recent years.

To those in the Western world, the extensive effort and resources which the Communist nations put into espionage and counter-intelligence is something of a mystery. We can understand the payoff in the recruitment of men who years later reach the positions of a Philby or a Burgess. But we are at a loss to understand a system which puts so much effort into the covert collection of relatively routine data, the great portion of which is often freely available by open means. For the fact is that a considerable part of the Communist espionage effort probably is superfluous from our viewpoint and is concerned with checking through covert means the veracity and completeness of information which is obtainable openly. This practice probably is a consequence both of the secretive nature of the Communist state and of a distrust of open sources as inherently subject to manipulation by the state or vested interests—as they are in fact in dictatorial states. What is important for our purposes is to observe that, because this is so, there is probably some inclination in these nations to give greater weight to information obtained clandestinely than to information obtained openly, even when the latter would seem to be of greater authenticity. This is not to say that the report of the KGB operative will necessarily carry more weight in Moscow than an authoritative article in the *New York Times* but only that there will be a predilection in favor of the covert report, other things being equal. In any event, the Communist system beyond any question is engaged in a continual, relentless, expensive and complex effort to ferret out our every secret, however minor, which might affect their national security.

And what does this have to do with warning? Its relevance to warning derives from the fact that the espionage and security services are a mirror which reflects the objectives and requirements of the national leadership on major foreign issues. The standing collection requirements for the espionage services, if we can be fortunate enough to obtain them, will provide us a blueprint of what our enemies most wish to know about us, which can often be very revealing. Moreover, a crisis almost invariably will result in some emergency collection requirements which can help us to perceive, at least in some degree, whether the enemy is primarily concerned that we may be preparing some action against him or whether he is preparing to initiate something. And, over the longer term, changes in the types of information sought may provide quite clear insight into enemy planning.

This point may be better illustrated by citing some specific examples from the fascinating accounts of the British penetration of the German espionage operations in the United Kingdom during World War II, which have recently been declassified and published. We will draw particularly on Sir John Masterman's discussion of what could be inferred concerning German war plans from the types of questions which were sent to their agents in

Great Britain—all of whom were actually operating under British control. He says:

> The most interesting point with regard to the traffic up to the beginning of 1942 is the evidence which it gives of enemy intentions. . . . In retrospect it is perfectly clear, even if it was not quite clear at the time, that enemy intentions could be gauged from the traffic of our agents with very, fair accuracy. In R.A.F. matters, for example, the majority of questions with regard to aerodromes was concerned all through the Battle of Britain with the position and defences of *fighter* aerodromes. Conversely, in 1941, when the British air offensive on the Continent started, interest swung over to *bomber* aerodromes and the landing grounds from which bombers operated. The extent of the danger to this country of invasion from Germany is naturally clearly mirrored in the messages.[2]

The queries to German agents on other topics also enabled the British to reach a judgment by late 1941 that the Germans had, at least for the time being, abandoned plans for any large-scale offensive operations against the United Kingdom. By 1944, when the strategic picture had materially changed and it was the Allies who were on the offensive, the nature of the Nazi questions again provided insight into German military planning. Masterman concludes:

> We should restate a conviction which established itself more and more firmly in our minds—viz. that a careful and intelligent study of all the traffic could and would have given an accurate picture of all the more important German interests and intentions throughout the war. In retrospect it is clear that more use could have been made of this product of double-cross agents' work.[3]

Possibly the most dramatic evidence cited by Masterman of the potential value of such agent queries for warning concerns a questionnaire given to a key German agent operating in Britain who was detailed on a special mission to establish an espionage network in the United States. On 19 August 1941, the British read and transmitted to the FBI a three-page German questionnaire of desired information on the United States, of which one-third was concerned with Pearl Harbor. It was further noted that, whereas most of the questions about the US were fairly general, those about Hawaii and Pearl Harbor were highly specific and called for details and sketches of airfields, hangars, bomb-depots, POL installations, and so forth. The logical inference, as Masterman points out, was that, if the US were to be at war, Pearl Harbor "would be the first point to be attacked, and that plans for this attack had reached an advanced state by August 1941."[4]

Although the nature of warfare may have changed since Pearl Harbor, the objectives of foreign espionage services have not changed greatly. All are concerned with the preparedness measures and vulnerabilities of the enemy. Moreover, there remains a high probability that there would be some modifications in the types of information desired if the threat of war appeared to be rising, and in the urgency of collecting and reporting this information. Changes in the structure and operations of the espionage services themselves also probably would be undertaken—such as a much greater reliance on undercover agents in expectation of a break in diplomatic relations, preparations for a change in means of reporting when diplomatic pouches and embassy radios cease to be available, and so forth. We should also expect an increase in those types of subversive activities which would directly further the war effort—such as the infiltration or surfacing of saboteurs and of experts in partisan warfare and the dissemination of false rumors. Obviously, no one could predict in advance exactly what changes in the espionage services and their means of operation would be undertaken, still less how much we might be able to learn of it. It also cannot be predicted whether acts of sabotage or political assassinations might actually be undertaken before hostilities began. There is no doubt, however, that the Communist espionage services all have their war plans and that some changes in operation would be undertaken before war broke out if time permitted. Thus, these activities are a potentially highly valuable, and perhaps unique, source of indications intelligence.

Problems of Compartmentalization

The tight security on espionage and counter-intelligence operations and the compartmentalization of those portions of the intelligence process which are concerned with these activities poses a potentially grave bureaucratic problem. The relative freedom of exchange of information among those engaged in the production of positive intelligence breaks down almost completely—at least at the analytic level—when counter-intelligence operations become involved. Even valuable positive intelligence derived from these operations may be bottled up or delayed in dissemination, often not for arbitrary reasons but on valid security grounds. And those aspects which are operational—such as surveillance of foreign agents and the nature of their contacts—are virtually never made available to the other side of the house at the working level, and rarely even at the highest levels of intelligence and within policy councils.

Thus the dangers that all relevant information may not be brought together in a meaningful pattern, and that indications will be "lost," are particularly acute in this field. Even greater than the separation of intelligence and policy and the compartmentalization of operational plans is the secrecy which surrounds counter-intelligence. This statement is not to criticize this policy—which is clearly essential in many cases, and sometimes a matter of life and death—but merely to note the potential seriousness of this situation for warning. Thus it could be highly important, *before* hostilities begin, that steps be taken to ensure that valuable warning information derived from counter-intelligence operations is integrated with other positive intelligence. Any procedures devised almost certainly would involve only a limited number of people but hopefully would attempt to see that some analytic group, however small, was coping with all (or at least nearly all) the pieces of the puzzle.

If it is any consolation to us, we might observe that our potential enemies might have much the same types of problems, in that their counter-intelligence operations also are highly compartmented, and that there is a much more restricted exchange of information in general. The involvement of the Communist leadership in the planning of major espionage, counter-intelligence and deception operations, however, does tend to ensure that the highest authorities will be cognizant of major developments as they occur. The chief of the KGB reports directly to the Politburo, and the present chief is a candidate member of that supreme decision-making body. The idea which sometimes prevails in Western democracies—that the chief of state and the foreign office should not become involved in or perhaps even be cognizant of "dirty tricks"—is not a problem which troubles the Communist world.

Notes

1. Some understanding of the number of personnel which may be involved in a major security operation in the USSR may be derived from a report that 20,000 military and civilian personnel were involved in security arrangements in Leningrad alone for President Nixon's visit there in May 1972—a report which the US Embassy in Moscow found "easy to believe."
2. J.C. Masterman, *The Double-Cross-System in the War of 1939 to 1945,* (New Haven and London, Yale University Press, 1972), 76.
3. Ibid, p. 177.
4. Ibid, pp. 79–80.

25

Warning from the Totality of Evidence

IN THE PRECEDING SECTIONS (Chapters 3 through 24) we have examined various types of military and civil preparations for hostilities, largely in isolation from one another. Obviously, in real life these various developments will not be occurring separately but in conjunction or simultaneously. Moreover, they should relate to each other in some more or less logical fashion if in fact a nation is preparing for hostilities. For example: there will not be urgent and massive civil defense preparations without various military preparations to bring the armed forces to higher readiness; there will not be political indications that the leadership has directed the implementation of certain wartime legislation without other evidence of mobilization. There will not be just military or just political indications, but a variety of developments in both fields which at least to some extent will be consistent or mutually supporting.

At the same time, however, there will likely be some inconsistencies in our evidence or at least gaps in our knowledge which will make us uncertain as to the significance of some of the developments, or of their relation to each other. We will not be sure what weight we should accord to any particular indication, or even to a large number of them collectively. No two situations will be just alike, and we cannot rely solely on precedent or history (although they may assist us) in coming to our judgments.

The Relative Weight of Political and Military Factors

At the risk of an oversimplification of this problem, we may note certain generally valid precepts.

First, political indications alone—in the absence of any significant military preparations or without the capability to act—are not credible and we will virtually always be correct in dismissing them as so much bombast or propaganda. For years, Communist China had a propensity for reserving some of its most violent propaganda for situations halfway across the globe in which it had absolutely no capability to act—e.g., Lebanon in 1958. In the years following its decisive defeat by Israel in 1967, Egypt's repeated calls for the recovery of its former territory carried little weight in the clear absence of a capability to defeat the Israelis in the Sinai. Similarly, the anti-American propaganda put out by North Korea over a period of years has been so intense and vitriolic that it has been meaningless as an indication of an intention to take military action within any foreseeable time period. We must always remember, however, that the national attitudes reflected in such propaganda are significant, and that such bitter hostility will make the military preparations (if or when they occur) potentially more meaningful and dangerous than might otherwise be the case.

At the other extreme, military indications alone—in the absence of any signs of political crisis or a deterioration in the international situation—also will tend to lose credibility. In such circumstances, we will be inclined to regard even quite extensive unusual military activity as an exercise or test of some kind, rather than a bona fide preparation for early military action. For example, a partial mobilization, which in time of political crisis would cause grave concern, would probably be dismissed as only an exercise in a period of political calm. In the absence of any crisis, even a highly unusual and potentially very ominous development may not cause much alarm; it will rather be regarded as a mistake of some kind, or an error in reporting, as in fact it often is. Whereas in a crisis such a development would likely be assessed as even more ominous than the fact alone might warrant, it will probably require quite a number of unusual military developments to disturb our complacency if we see no positive political indications. Although this is in part a psychological phenomenon, it is also historically valid. Very few wars have started without some deterioration in the political situation, or some development which would increase the possibilities that a nation might decide to launch military operations.

There is, however, some limit to the number of major military preparations which may be undertaken in a period of political calm without arousing concern. Obviously, this would be particularly true if one of our most powerful potential enemies were to begin extensive and unusual military preparations, even though the political atmosphere was relatively "friendly." The idea, advanced by some, that the USSR could mobilize and redeploy its forces against NATO in a period of calm and convince us that it was just "an exercise" begins to strain credibility. There would be some point in that process, however complacent we might be at its start, that the sheer buildup of capability would cause grave concern and almost certainly some type of military preparations on our part. This does not necessarily mean that we would reach a positive judgment that Soviet attack was likely, but we would come to appreciate that we could no longer say with confidence that it was very unlikely.

In real life, we rarely see the situation in which political and military indications are totally out of phase or contradictory. Each will be contributing, in varying measure perhaps, to our assessment of the enemy's likely course of action. It has been observed that, in normal times, we will usually give somewhat greater weight to political indications than to military developments—this reflects our general sense of the attitudes and intentions of our adversaries, usually borne out by many years of experience. It is also essentially our national estimate—that they are not going to go to war without some reason, and that we will have some indication that the situation has changed before they would take such a decision. On the other hand, once the situation has changed and the political atmosphere is deteriorating, we will probably give greater weight in the crisis situation to the military indications as our best guideline to the enemy's intentions. This in turn reflects two historically valid principles: political indications can be ambiguous or even misleading, particularly if the adversary is seeking to confuse or deceive us; and the extraordinary buildup of military capability is likely to be the best single indication of the enemy's course of action, a point made several times previously in this work.

Isolating the Critical Facts and Indications

Individuals lacking experience with real warning situations nearly always have considerable misconception about the nature and quantities of information which are likely to be received, and the problem of interpreting it. Whereas the inexperienced tend to believe that warning "failures" arise from totally inadequate information ("we didn't have warning"), experienced analysts have learned that the reverse may be the case—there is almost too much information, too many reports, too many military preparations, too much "warning." It must be conceded that this is not always the case, and that there have certainly been areas and circumstances in which our information was very inadequate. A review of the evidence available prior to the outbreak of most recent conflicts, however, will show that a great deal of information was usually available. What was lacking was probably the evidence of the final decision to go and the evidence of the final military preparations which would have given a clue to the timing of the attack—problems which we will discuss in coming chapters.

In any large volume of political and military reports or indications, some obviously will be of far greater importance than others for the judgment of the enemy's intentions. In the preceding chapters, a number of such critical facts and indications have been discussed, particularly highly unusual military developments which can be expected to occur only in preparation for combat. As we also have noted, many political and civil indications may be much more ambiguous, but some will be much more meaningful than others for warning, and hence should be accorded much more weight in the assessment of the enemy's intent. It will therefore be important that these particular meaningful preparations be singled out and accorded the attention they deserve. The question should not be simply, is this a likely preparation for war? There will probably be a great many developments in this category. The crucial question may be, how rare is it? How often has it occurred at all in peacetime, including crises which did not lead to conflict? How likely would it be to occur except in preparation for war? If the answers show that even a few critical or nearly, unique indications are showing up, the odds of course are materially increased that the nation in question is preparing for and will probably initiate hostilities. The more advanced and sophisticated the military forces and the economy of a country are, the more such distinctive preparations will be required for war. Preparations for nuclear war would involve an unprecedented range of activities, some of which would probably never be seen except in preparation for that contingency. It follows, therefore, that:

All Indicators Are Not Ambiguous

A great disservice has been done the community and the warning system by some rather casual statements that "all indicators are ambiguous." Such comments are not dissimilar in lack of perception to the claim that "We can judge the enemy's capabilities but we cannot judge his intentions" (see Chapter 5).

Those who make such off-hand judgments are probably familiar neither with the examples which can be drawn from history nor with the specificity of some items on indicator lists. Or—which may be equally likely—they are using the word "ambiguous" in a highly ambiguous sense.

It is probably true that there is only one totally reliable, unequivocal indication of an intention to attack—and that is instantaneous access to the enemy's decision to do so and/or the order to implement it. Even where total preparation for war has been accomplished, where all military indications are "positive," and even when the political decision has already been made in principle to attack, there is always the possibility that the leaders will change their minds or that some last minute event will cause them to postpone or to call off the operation entirely. In this sense, it may be said that all indications but the one are subject to some measure of doubt or uncertainty and can never be viewed as absolutely conclusive evidence of the enemy's intent.

But there are, as emphasized in the preceding discussion, a number of military indications which are not in themselves ambiguous. That is, they are the steps which are undertaken only in preparation for hostilities, which virtually never occur in peacetime, which are not just "more of the same" but different from what goes on from day to day. They do not occur for exercises, they do not occur (or only to a very limited extent) in practice mobilizations or other drills. They are the developments which truly distinguish war from peace and which, in the Soviet Union, we have never seen, at least since World War II. They are the manifestations of the implementation of the war plan, and they include such developments as: full national mobilization; the institution of full combat readiness in all military forces; the formation of wartime commands; the release of nuclear weapons to the authority of the commander; and a number of other similar although less dramatic measures.

There are further a number of lesser military developments which, although not necessarily indicative of imminent hostilities, are positive indications that the combat readiness and capabilities of forces are being raised, or that they are being deployed into positions for attack. To call these measures "ambiguous" is highly misleading, for the military measures themselves are not. They are not exercises but bona fide measures to raise the combat capabilities and readiness of forces for a particular action. Even if that action is not finally implemented, the preparedness measures themselves should not be dismissed as of doubtful or ambiguous significance. Many of the measures taken by Soviet and Warsaw Pact forces prior to the invasion of Czechoslovakia were in this category—they materially and obviously raised the capabilities of these forces for such an operation and bore no resemblance to normal "exercises." They brought these forces to a very high degree of readiness to invade—a fact which the intelligence community recognized and stated. If the situation was "ambiguous," it was only because firm evidence was lacking (or many chose not to accept as likely) that the Soviet Union finally would go through with the invasion. But this did not negate the validity and non-ambiguity of the military developments themselves.

Negative Indications and Problems of Concealment

In assessing the enemy's intentions, it is necessary not only to take note of what he has done, but also of what he has *not* done. If we can determine for sure that he has not taken certain essential preparations for conflict, or even has taken some which might reduce his readiness for combat (such as releasing seasoned troops), this will materially influence our conclusions. In some cases, knowing what has not occurred can be the most important factor of all.

Unfortunately, it is often very difficult to find out that something has not happened. This is particularly true of the whole range of preparations, both military and civil, which are not readily discernible or which involve relatively little overt activity. There are other preparations, particularly those involving major deployments or changes in the normal patterns of military activity, on which we often can make a judgment with some degree of confidence that certain things either have or have not occurred.

In compiling a list of what is often called "positive" and "negative" indications, therefore, great care should be taken to distinguish true negative indications (things that we expect to happen prior to hostilities but which have not) from just plain lack of information. In some cases, a large portion of the seeming negative indications will turn out to be in the no information category. On some of these, we may be able with sufficient collection to make a determination one way or the other. On many others, however, our chances of finding out anything are poor, and sometimes very poor. We must be careful not to mislead our consumers into believing that we know more than we do, and it may be necessary to point this out quite explicitly. The indications or current analyst should avoid phrases such as "we have no evidence that" when the chances of getting the evidence are poor, and he should not otherwise imply in any way that the information he is presenting represents the sum total of what the enemy is up to. It may be helpful just to compile a list of the things that logically could or might have happened which we cannot tell about one way or the other. The consumer of intelligence in his turn must have a realistic understanding of indications intelligence and our collection capabilities lest he equate a lack of reporting with lack of occurrence.

Reporting from field collectors also should be geared to ensure in a crisis situation that those at headquarters know what the collector has covered or even can cover, when he files his "negative" report or fails to send a report at all. A true "negative indication" from the attaché is not the absence of a cable, from which we assume that all is well, or even the report which reads: "Troop movements, negative; mobilization, negative." We may need to know what parts of the country he and his colleagues have covered, and whether any troop induction stations or reserve depots have been reconnoitered to be sure what "negative" means.

Subject to these provisions, the careful compiling and reporting of true negative indications can be a most important portion of the totality of evidence and hence of the final judgment of the enemy's intentions.

Urgency

A distinguishing feature of most crises which result in hostilities and of the preparedness measures which accompany them is urgency. There is an atmosphere which surrounds the bona fide pre-war situation which differentiates it from exercises, shows of force, or even political pressure tactics. Although it is somewhat difficult to define this atmosphere, or to explain exactly what makes it seem "real," an important ingredient nearly always is urgency. This sense that there is a race against time that things are being done on an accelerated schedule, that the pressure is on is likely to be conveyed to us in a variety of ways. It usually will affect both military and political activities and be evident in a number of anomalies or indications that plans have been changed, trips cut short, exercises cancelled, propaganda changed abruptly, and so forth. Only in rare instances—and those usually where our collection is poorest—do we fail to obtain some evidence of this urgency. Where the pace is leisurely and there appears to be no deadline for completion of the activity, we will usually be correct in judging that it represents a long-term or gradual buildup of capabilities rather than preparation for early hostilities. The general absence of urgency or hurried preparation has been one of the major differences, for example, between the Soviet military buildup along the Chinese border over a period of years, and the precipitate movement of forces prior to the invasion of Czechoslovakia.

One note of caution is in order, however. There are instances of long preplanned and deliberate attack—the North Korean attack on South Korea in June 1950 is a prime example—in which evidence of urgency or even any particular sign of crisis at all may be lacking. Where a nation has more or less unlimited time to prepare and is practicing a deliberate political deception campaign designed to lull the adversary, it may under favorable circumstances be successful in concealing or suppressing any signs of urgency. (See further discussion in the next chapter.) In the case of the North Korean attack, our very limited collection capabilities undoubtedly also contributed heavily to the surprise we were not even alerted to the possibility of the attack when it occurred.

It must also be said that urgency of activity alone, of course, is not a firm indication of intent to undertake offensive operations, since obviously there may be circumstances calling for speedy military preparations and hurried political decisions when there is no hostile intent. Even in this case, however, the urgency of the activity will usually indicate that the nation is genuinely concerned, that it regards the threat seriously, and/or that it is not bluffing.

Some Guidelines for Assessing the Meaning of the Evidence

Crises are marked by confusion, by too much raw information and too little time to deal with it, by too many demands on the analyst and so forth. It would be nice to have lots of time for the interested and knowledgeable analysts to assemble and review their evidence, make their arguments, reexamine the facts, and revise their judgments and conclusions, much in the laborious fashion that national estimates are prepared. In warning, unfortunately, time often does not permit this and it frequently does not even permit some of the less time-consuming means of getting analysts together to discuss the material and exchange views on what it all means.

In these circumstances, analysts and consumers alike may profit from some relatively simple guidelines designed to assist in evaluating the evidence and the intention of the enemy.

We begin by assuming that the enemy is behaving rationally and that he is following some logical and relatively consistent pattern of action in achieving his objectives. Although this may not always be the case (nations as well as individuals have sometimes acted irrationally and inconsistently), it is well to start with the logical analysis of the enemy's behavior before assuming that he may act irrationally. As a result, we also assume that war is not an end in itself for him and that he will not resort to hostilities so long as there is some reasonable chance of achieving his objectives by means short of war. We therefore start with the five following questions designed to clarify our own thinking about what the adversary is up to. The questions are:

1. Is the national leadership committed to the achievement of the objective in question, whatever it may be? Is it a matter of national priority, something the leadership appears determined to accomplish?
2. Is the objective potentially attainable, or the situation potentially soluble, by military means, at least to some degree?
3. Does the military capability already exist, or is it being built up to a point that military action is now feasible and victory likely to be attainable? Or, more explicitly, does the scale of the military buildup meet doctrinal criteria for offensive action?

4. Have all reasonable options, other than military, apparently been exhausted or appear unlikely to have any success in achieving the objective? Or, more simply, have the political options run out?
5. Is the risk factor low, or at least tolerable?

If the answer to all the questions is a firm yes, logic would dictate that the chances of military action are high. If the answer to any one of them is no, then it would appear less likely, or even unlikely, that the nation will resort to military action now, although of course circumstances might change so that it would decide to do so in the future. If two or three answers are no, the chances of military action would logically appear to drop drastically to the point of highly improbable if four or all answers are negative.

Applied to some recent indications problems, this technique yields some interesting results. For the Soviet invasion of Czechoslovakia, the answer to all five questions is yes—although some persons might maintain that Soviet political options were not entirely exhausted by 20 August 1968, a series of political measures had failed to bring the situation under control, and there was little reason to believe that more such pressures would succeed. For the Arab-Israeli conflict of June 1967, the answers from Israel's standpoint also are yes to all five, although slightly less clearly or categorically perhaps than for the USSR in 1968, i.e., the risk factor was seemingly a little higher and the exhaustion of political solutions perhaps a little less certain. For Egypt in 1971, the answers to questions one and two are emphatically yes, to four also yes (from a realistic standpoint), but the likelihood of military action drops drastically because Egypt lacked the capability for successful military action and the risk factor was high for India in the India-Pakistan war of December 1971, so the answer to all five questions again is yes.

In the Sino-Soviet border controversy (which reached its most critical point in 1969), we can come to a firm yes only on question one—the Soviet leadership did appear committed to "doing something" about the China problem, particularly after the Damanskiy Island incident in March. To all other questions on this thorny problem, however, the answer is either no or at least uncertain. It was highly doubtful that a Soviet military attack would have "solved" or even lessened the China problem. The Soviet Union could not build up sufficient military force actually to conquer the Chinese people in war—except possibly by the use of nuclear weapons. The employment of these, in turn, would make the risk factor very high—both militarily and politically. And finally, difficult as the Chinese might be to negotiate with, the political options had run out. And in due course talks—not very fruitful but still talks—were begun, and the crisis atmosphere which had prevailed began to abate.

Because of the different nature of the Soviet actions in Cuba in 1962 (obviously, the Soviet Union never intended to go to war over Cuba), the foregoing questions cannot all be literally applied to the Cuban missile crisis. Insofar as they are applicable, however, the answers do not yield a positive yes which would have made the Soviet action logically predictable or consistent with previous Soviet behavior. In particular, the risk factor—from our standpoint and in fact—was extremely high, and the Soviet action is explainable only as a gross miscalculation of what the US reaction was likely to be.

Thus these questions, although useful as a logical starting basis for the examination of the meaning of our evidence, are not a foolproof guide to an assessment of the enemy's intentions. For there will also be the cases in which the adversary's action will not necessarily be logical—where he may resort to military action, even though the answer to one or even more of the five questions is no. For a variety of reasons—miscalculation of the opponent's strength or reaction, overestimation of one's own strength, frustration, internal domestic pressures, patriotic hysteria, revenge, a fit of pique, or just plain desperation—a nation's leadership may decide on imprudent or even disastrous courses of military, action which are clearly not in its national interest.

Nearly all conflicts are final acts of desperation when other means of solution have failed. In many cases, the instigator of the military action nonetheless has followed a rational and consistent course of action, and after due deliberation and after all other options have failed to yield results—has decided on military action as the only method which will achieve the desired result. Military solutions are not inherently irrational acts, particularly if they are likely to succeed. The Soviet invasion of Czechoslovakia, for example, was a carefully deliberated, meticulously planned, coldly rational, and entirely logical course of action; although there were political (but not military) risks, they were far less from the Soviet standpoint than permitting Czechoslovakia to pass from control of the Communists.

Before we conclude that some other nation is acting "irrationally" in going to war, we should carefully examine our own attitudes and make sure we are not rejecting such action as illogical because we either do not fully appreciate how strongly the other country feels about it, or because we are just opposed to war on principle as an instrument of national policy.

I believe that the systematic application of the method described above will far more often than not yield positive and correct results. At a minimum, it is a method of helping ourselves to think objectively about the evidence as a whole and to avoid, insofar as possible, substituting our own views for those of the other guy.

But there will remain those cases, like Cuba in 1962, which are not logical and do not meet objective criteria for rational action. It is this imponderable, of course, which so vastly complicates warning. We must allow for those cases where the risk factor is high or where military action is not likely to solve the problem and may even be potentially suicidal. When there is good reason to suspect that the leadership of the nation in question may be acting irrationally, the two most important questions are slightly modified versions of one and three:

- Is the national leadership so committed to the achievement of the objective, or so obsessed with the problem, that it may even act illogically in an effort to achieve its goals? and
- Is the military capability being built up to the maximum possible for this action, even though the chance of success is doubtful?

26

The Impact on Warning of Circumstances Leading to War

THE VARYING CIRCUMSTANCES under which wars may start, and the differing motivations or objectives of the nations which begin them, inevitably will have considerable effect on the indications of the coming of war and our assessment of them. The causes of war and the reasons why nations resort to conflict are, of course, enormously complex, and it would be absurd in a single chapter to attempt even to outline the scope of this problem. Our purpose here is not to analyze the causes of wars, but only to describe how warning is affected by some of the various circumstances surrounding the outbreak of conflicts. In particular, we are concerned with such things as deliberately planned aggression versus wars which come about because of some change in circumstances, or by miscalculation or escalation.

The Deliberately Planned Aggression

Historically, many and possibly most wars have started from territorial ambition or a simple desire to gain power, conquer other peoples, or even to rule the world. In these circumstances, the quest for power *is* the cause of conflict, and war is not brought about by any circumstance other than the ambitions of rulers, such as Genghis Khan, Napoleon, or Hitler. Although such stark militaristic expansionism has not in the past quarter century been a cause of many conflicts, it was the basic cause of World War II both in Europe and the Pacific. Thus, those who correctly perceived the coming of World War II were those who recognized that both Hitler and the Japanese militarists were bent on conquest and that war was an instrument of their national policies rather than something to be avoided. In the broadest sense, *the* warning of the coming of the war was the recognition of this. Beyond this, warning became a question of when and where conflict would break out, rather than whether.

The militarist who is determined on conquest and has the requisite power to initiate the conflict obviously has many advantages in the planning of his operations. Most important is that he controls the coming of the war, which will be initiated by and large at the time and place of his choosing. He need not be pushed into war until he is ready or make his military preparations in haste. He can try military blackmail to secure his ends and if successful obtain some of his objectives without resort to war, if he chooses to operate this way. Or he can, at least theoretically, avoid any kind of ultimatum or demand on his intended victim and attempt to launch a surprise attack without any "political warning" or seeming deterioration in their relationship.

These options open to the military aggressor also affect the type and number of indications which may become apparent to the adversary or be concealed from him. As is most evident, the political indications can conceivably range from virtually nil (no hostile propaganda, no diplomatic moves or threats, wholly successful concealment of the decision to attack, etc.) to the most obvious kinds of political blackmail, ultimatums or overt declarations of intent to attack. In practice, there have been very few instances in recent times of attempted total political concealment or absence of crisis, although, as noted in the preceding chapter, the North Korean attack on South Korea in June 1950 comes close to it. The point is that, in closed societies and with the practice of maximum security, the potential aggressor at least has a presumed capability of achieving political "surprise" of this type. At the least, there is always the chance that he might try it.

The long-planned, deliberate aggression also permits considerably greater concealment of many military preparations, primarily because security and deception measures can be most carefully planned and implemented. Many steps also can be undertaken more gradually, so that there is less discernible disruption of normal military and civilian activity. In contrast, sudden, unexpected crises requiring precipitate and unplanned moves of military forces in response to the emergency nearly always are apparent in some degree in nations where we have any significant collection capability. They often are accompanied by breakdowns in military security which would be almost unheard of in normal circumstances.

On the other hand, the instigator of deliberate aggression normally will initiate his military preparations much earlier and usually more extensively than in the unexpected situation in which war comes about because of some external change in the situation. Thus, although the preparations individually will be less obvious (or not obvious at all), the collection services may have much more time in which to detect them, and there will be more readiness measures to detect, than in the crises. The analytic elements of the intelligence services in turn will have more time in which to make their evaluations, recheck their sources, and so forth. In general, given the serious problem of delay in the acquisition of confirmatory data from within most closed societies, time is likely to be on the side of the victim in the long-term buildup. That is, by and large his chances of detecting the military preparations will be better the longer he has to do so, even though the security measures to conceal them are also better. At the same time, however, the preparations may seem to lack an urgency which can be deceiving, and the intended victim of the aggression may be more inclined to view the preparations of his enemy as a long-term buildup of capabilities or contingency preparations rather than as indications of an intention to attack. Obviously, each case will be different in some degree, and it would be misleading to attempt any generalizations on this subject.

The War Brought on by Changes in Circumstances

Within recent years, the deliberately planned aggression has been much less frequent than the war which comes about because of some external development which alters, or threatens to alter, the balance of power, or because of some other change which worsens a long smoldering situation. The variety of such developments is considerable. They include sudden spontaneous disturbances or outbursts (such as the Hungarian revolt in 1956), nationalist or chauvinistic actions which threaten the interests of another power (such as the Egyptian seizure of the Suez Canal in 1956 and the Egyptian closure of the Gulf of Aqaba to Israeli shipping in 1967), an escalation resulting from border incidents (such as the Sino-Indian conflict in 1962), to name a few. Potential causes of conflict also include provocative actions or threats of actions designed primarily to test out the opponent and see how much he will put up with, such as the Chinese Communist shelling of the Nationalist-held offshore islands in 1958, or the series of Soviet threats to Berlin from 1958 to 1961. Other examples could be cited.

The point is that situations of this type, whatever precipitates them, are obvious occurrences which demonstrably raise the international temperature and in which the threat of possible conflict is nearly always immediately apparent. And where hostilities do follow, it may not be by choice of the instigator but because he feels compelled by circumstances to resort to force or the situation simply gets out of control. Some nations, of course, will be less unwilling than others to resort to force, but nonetheless the situation is not entirely of their choosing. The conflict, if it comes, arises at least in part from circumstances or miscalculations rather than machiavellian design.

The developments which precede such conflicts, or potential conflicts, are likely to vary to some extent and sometimes considerably from those which precede the deliberately planned aggression. This difference will not be primarily in the actual preparations which precede the outbreak of the conflict; in fact, the military steps may be virtually identical, and some of the same types of civil preparation will likely be taking place as well. The differences arise rather from such factors as: the motivations of each of the participants to the dispute; their willingness or reluctance to resort to conflict; their readiness to negotiate or to seek a genuine compromise solution; the actual status of negotiations if they are begun; the effectiveness of the intelligence services in ascertaining the level of preparedness of the antagonist; perceptions of the intentions of the other party regardless of the effectiveness of intelligence; assessments of whether time is or is not on their side; respective estimates of relative military capabilities (not always accurate); military doctrine on preemption and surprise; pressures from other nations to hold off operations or to limit their scope, and so forth. The foregoing are only some of the complex factors which can influence whether the leadership of a nation will resort to conflict—which it may not wish, would not have chosen in other circumstances, and might have avoided had it better understood the adversary or not misinterpreted his preparations.

It is obvious that the forecasting of hostilities in such complex circumstances can be a highly hazardous occupation and potentially fraught with difficulties which do not arise in the case of the nation which is firmly committed to conquest and has the clear military capability to achieve its ends by force. It is impossible to obtain conclusive evidence of the intentions of a nation which has not yet made up its own mind what to do. In these circumstances, the finest penetration of the highest councils of government will not provide definitive answers but only information that the decision has not yet been made. Such access, however, and even intelligence of lesser quality, may provide us understanding of the options as the leadership sees them and a perception of the circumstances which might in the future result in the firm decision to go to war. It thus may permit us to make quite a good judgment of the probabilities of conflict even though the decision has not yet been made. But in the best of circumstances there will likely be a considerable element of uncertainty as to whether the conflict will or will not be avoidable and, if not, when it may finally be precipitated.

Given all these uncertainties, it must be said that our record in perceiving the coming of hostilities in these complex situations is perhaps better than might be expected and would seem to compare favorably with our record in predicting the deliberately planned aggression. Thus our recognition of the likelihood of conflict between Israel and Egypt in May–June 1967 was considerably better than our perception that North Korea would attack South Korea in 1950, and forecasts of the Indian-Pakistani conflict in December 1971 were more forthright and accurate than forecasts that India would move against the little enclave of Goa in December 1961.

It would no doubt be highly misleading to attempt to generalize why this is so. But a primary reason unquestionably is that a crisis generates a recognition that a conflict may ensue and an understanding of why it may occur. Both sides usually will take pains to make their concern evident, and many if not most military preparations may be poorly concealed. Collection is stepped up, and developments which would pass virtually unnoticed in a period of calm are promptly reported and disseminated; they may, in fact, be given more weight than is warranted because of the charged atmosphere. The press and other public media are filled with material on the crisis. The French have a word for it—intoxication. Thus the coming of the conflict, even when the intelligence services fail to predict it with certainty or expect it too soon, cannot really be said to be a surprise.

The One-Sided Versus the Two-Sided Buildup

Regardless of the reasons for conflict and the motivations of the attacking nation, our interpretation of the military moves and issuance of warning will be considerably complicated if both sides are preparing for war and reacting to the preparedness measures of the other. Where the two powers are relatively equal in military capabilities, and both are building up their forces, the interaction of their preparations may make it difficult to tell which nation, if either, is the potential aggressor and which is only preparing to defend itself. Obviously, the chances for miscalculation and misjudgment are compounded, and an inadvertent border violation or minor incident potentially may trigger an outbreak of major hostilities in which neither side appears as the clear-cut aggressor. The problems of war by miscalculation rather than design have increasingly concerned both the intelligence community and policy makers in recent years. Some aspects of these problems from the standpoint of warning will be considered in the remainder of this chapter.

First, it should be noted that the one-sided buildup of military force—or the instance in which one nation to the controversy has such overwhelming power at its disposal that it would be ridiculous to assume that its adversary could possibly initiate the conflict—are by no means infrequent. In a number of conflicts since World War II, one side has had a great preponderance of power and has been able to employ force—or to threaten to employ it—with relative impunity. Indeed, it is the clear preponderance of force which encourages the military solution rather than extended and possibly fruitless negotiation, while, conversely, the threat of neither side was "inevitably" forced into the conflict, which might have been avoided by a genuine desire to negotiate or compromise in the interests of avoiding war.

A conspicuous example in which gross miscalculation threatened to lead to war, but in which conflict was avoided, was the Cuban missile crisis. Khrushchev's shipment of strategic missiles to Cuba—which must rate as one of the most astonishing and dangerous misjudgments ever made, certainly by the usually cautious Soviet leaders—is a prime example of the type of miscalculation which could precipitate a conflict between major powers. Since both nations, in this instance, were above all anxious to avoid war, and the potential for escalation to nuclear conflict was both real and terrifying, the danger was averted. For analyst and policy maker alike, however, the missile crisis must serve as a constant reminder that miscalculation of the effects of military actions is a greater danger in the age of nuclear weapons than it ever has been before.

It will be apparent that such miscalculations present extraordinarily difficult problems for warning analysis. The prediction of seemingly irrational behavior, particularly in the absence of strong supporting evidence, is a virtual impossibility for the intelligence system—except as a contingency warning to the policy maker of a *possibility* which might be developing, or a threat which might be in the making. As is well known, a special national estimate on Cuba, issued on 19 September 1962, reached the conclusion that the introduction of Soviet strategic missiles into Cuba was unlikely—a judgment which correctly has been called both logical and wrong—taken into account a variety of political and military factors. Obviously, some access to the nature of policy decisions may be even more valuable in such instances than in the simpler situation. Lacking this, there are perhaps some questions which may help to clarify analysis, in addition to the five general guidelines in Chapter 25. To suggest a few:

- Are the parties really seeking a political solution or willing to compromise, or is either or both determined to solve the problem on its terms once and for all?
- Closely related, are great issues of national prestige involved, or even national survival of one of the parties?
- Has one or the other made public commitments, or taken other steps, which would be very difficult to reverse or revoke?

- Has either or both taken steps to justify military action, e.g. to its own people, its allies or the world at large?
- Does one of the countries have sufficient military advantage that it can reasonably expect victory if it attacks?
- Are the military preparations placing a heavy drain on the national resources of either or both sides, and hence unlikely to be sustained indefinitely without some action to resolve the crisis?
- Are the military preparations and deployments of either side essentially of an offensive nature, and are the defensive preparations consistent with an expectation of retaliation for a coming attack?

On this latter point, the reader is referred to Chapter 17 for a discussion of offensive versus defensive preparations. It must always be remembered that the most intensive and complete defensive preparations may be taken by the nation which is preparing to attack and expects early retaliatory action. The analyst must be careful not to write off such preparations as indicative of nothing more than fear that the other side may initiate the attack.

Problems of Miscalculation and Preemption

Much has been written on these subjects in recent years, and it may be well to attempt to define what is usually meant by the terms.

"War by miscalculation" usually defines the situation in which a provocation or relatively limited action by one party causes an unexpectedly strong response from the other, and in which a series of escalatory steps follows, resulting finally in the war which, in some instances, neither side presumably wanted. Although there are few live instances in recent years, the Arab-Israeli Six-Day War of June 1967 is probably one. In this case, war followed Egyptian miscalculation of the Israeli reaction, particularly to the closure of the Gulf of Aqaba, this misguided action on Cairo's part constituting the primary immediate cause of war. (It may be added that there is good reason to believe that the USSR, or elements of the Soviet intelligence services, also contributed to Cairo's miscalculation by misinforming Egypt concerning Israeli intentions, this error in turn having been a serious miscalculation somewhere in the Soviet apparatus.) On the other hand, there was no miscalculation by Israel in its attack, which was a carefully considered and superbly executed action. But protracted war or escalation of the conflict will tend to encourage negotiated solutions or the deferment of any solution. These elementary facts require no elaboration; they are the foundation of "balance of power" politics. Thus, the USSR can employ force against Hungary and Czechoslovakia with no fear of military retaliation, but it cannot do so against West Berlin. It is in fact constrained even from major military threats against Berlin, or the initiation of even minor military incidents on the autobahns, because of the grave dangers of escalation. Ultimately, when political threats, bluff, military maneuvers, and other measures short of direct military action had failed to achieve results, the USSR simply put Berlin on the back burner.

The situation marked by a unilateral buildup of force is the least complicated and should be the least difficult of warning problems, in that the military activity usually cannot be written off as "defensive," a reaction to preparations by the other side, or otherwise ambiguous. Normally, the only reasonable grounds for regarding such buildups as inconclusive indications of hostile intent are that: the military buildup is intended as pressure to force surrender or capitulation without having actually to employ it; or it is a contingency preparation for possible action at a later date, meaning the leadership has not yet firmly made up its mind and is "keeping its options open." (See discussion of this in the next chapter on the decision-making process.)

But what guidelines are there to help us when the situation is not so clear cut, when both sides are mobilizing and neither enjoys an overwhelming preponderance of power? It would be nice to have some pat answers, but of course there are none. Each situation will be unique and judgments of intentions will necessarily have to be taken into account. The evidence which might have supported a contrary conclusion was circumstantial and, while it clearly indicated that the Soviets were "up to something" extraordinary in Cuba which *could* include the introduction of strategic missiles, such a judgment was not then susceptible to proof. Thus, even those who believed such action likely—foremost of whom was the Director of Central Intelligence, John McCone—could do no more than to urge its consideration as a possibility or probability. For, without proof, the President could not act—he could perhaps have issued even stronger warnings to Moscow, but he could not have imposed a "quarantine" on Soviet ship movements to Cuba for produced convincing evidence for his action to present to the US public, the United Nations, or to Moscow. The Cuban case is a good example of the limitations of warning intelligence and a demonstration that there are times when only more collection can provide the answer and that the first duty of intelligence in such circumstance is to be sure that its collection systems are geared for a maximum effort (see discussion in Chapter 4).

Preemption or *preemptive attack* is defined in the *JCS Dictionary* as "An attack initiated on the basis of incontrovertible evidence that an enemy attack is imminent." It is to be distinguished from the longer term preventive war or preventive attack, which follows from a belief that war, although not imminent, is inevitable or at least highly probable, and that delay will entail greater dangers or risks than attacking now. It is difficult to find an example of true preemption. The Israeli attack of June 1967 lies somewhere between preventive and preemptive attack, since there was no clear indication that an Egyptian attack was imminent, let alone under way, initial Israeli claims to the contrary notwithstanding.

The appearance of nuclear weapons of course has vastly increased the dangers of preemption—and some believe also its likelihood, although this is arguable. The problem clearly is closely tied to that of miscalculation, for he who goes so far as to tempt preemption and he who in turn preempts may be guilty of the greatest miscalculations of all.

Defusing the Crisis Hot Lines and Other Devices

It is these acute problems—miscalculations resulting from inadequate communications, the danger of preemption based on false intelligence or misinterpretation, and other possibilities for gross misunderstandings between nations—which have led to the establishment of "hot lines" and other devices to facilitate communications between the heads of state or other high-ranking officials. No one could deny the merits of such systems for rapid and secure communications or their potential value for avoiding dangerous confrontations when both sides in fact are seeking to prevent misunderstandings and to avoid war. At the same time, it is doubtful that hot lines will serve to avert conflicts if one side is determined to continue on a collision course, and there is nothing that can be said by such means that cannot be equally well conveyed through more conventional channels, perhaps a little more slowly. There is also a grave danger that such links would serve as top-level deception channels in the event of a premeditated attack.

From the standpoint of the indications analyst, and indeed the intelligence system as a whole, the use of high-level direct communications between heads of state introduces still another method by which intelligence will likely be denied both pertinent facts and knowledge of policy decisions and moves by its own side. The gulf which today so often separates intelligence on the one hand and the policy and operational levels on the other is particularly dangerous in time of crisis (see discussion in Chapter 6). Without questioning the other advantages for the policy maker, the denial to intelligence of information from high-level private communications can only compound the ever-present dangers that intelligence will be issuing judgments which are less complete or accurate than they might be, or failing to report some information altogether because its importance or relevance was not apparent.

27

Reconstructing the Enemy's Decision-Making Process

If the final objective of warning analysis is the understanding of what the adversary is going to do, then the knowledge or recognition that he has decided to do something is the ultimate achievement. The highest goal of every espionage service is the penetration of the enemy's decision-making machinery—the hidden microphone in the conference room, or the agent with access to minutes of the conference, etc. To have this type of access is to be sure, or nearly sure, of the enemy's intentions, and will make superfluous a vast amount of information, however valuable in itself, from lesser or secondary sources.

Since we are most unlikely to have such access to the highest councils of our enemies—or if it could be obtained, it would be a highly vulnerable and perishable asset we must try to do the next best thing. We seek sources and information which will best permit us to deduce what may have been decided or to infer what the adversary's objectives and plans may be. In practice, in a crisis or warning situation, this will mean that we must examine virtually all the available evidence in an attempt to perceive what the pieces both individually and collectively may tell us about the enemy's decisions.

Obviously, this is both a highly sophisticated and very difficult analytic problem. It is also one of the most controversial aspects of the warning problem, on which there is apt to be the widest divergence of opinion in a live situation. Moreover, very few guidelines appear to have been devised to assist the analyst or the policy maker to follow some logical process in reconstructing the enemy's decision-making process. In the pressures of a crisis situation, and lacking any body of experience or agreed "rules" which might be of assistance, there has been some tendency in the intelligence community to ignore this problem. What should be of highest priority in the analytic process—the attempt to decide what the enemy has decided—is often shunted aside in favor of mere factual reporting of what is going on which is obviously much easier and less controversial. Too often, the reconstruction of the enemy's decisions and planning is attempted only after the crisis has been resolved, and thus becomes one more piece of retrospective or historical analysis, rather than something which might have helped us to foresee what was going to happen. Some brilliant post mortems have been produced, which have revealed that there is considerable talent for analyzing the decision-making process by inferential means. Such studies almost invariably also dig up pieces of information which were not considered at the time. But they nearly always are produced too late to help analysis in the current crisis and hence to be of any assistance to our own decision makers.

Clearly, it would be very useful to have some type of methodology which would help us to deal on a current basis with this elusive, but highly critical and sometimes decisive, factor in warning. It would be presumptuous to suggest that the remainder of this chapter is going to provide the answers, or some kind of simple and foolproof methodology. Its purpose rather is to assist the analyst to ask the right questions and to point out some of the more obvious aspects of this problem which have often been overlooked. After a brief discussion of these points, we will examine what we know about Soviet decision-making in two major crises.

Some Elementary Guidelines for Decision Analysis

a. Actions flow from decisions not decisions from actions. On the surface, this appears to be a truism, and almost an insult to the intelligence of the reader. Yet experience shows that this elementary principle is often not understood in crisis situations. In case after case, there has been a tendency to project into the future enemy decisions which must in fact already have been taken. The impression is left that the adversary is highly confused, hasn't decided anything yet, and is just doing things with no plan behind them. Thus, even major deployments of military forces may be downplayed or written off in such commonly used phrases as: "The deployment of these units significantly increases the enemy capability to attack, *if he should decide to do so.*" The last phrase is not just gratuitous, it can be downright misleading. It suggests that the forces are being moved without any plans in mind, that the adversary does not know yet what he is going to do with them, that major actions have been taken without any reason for them and that the enemy is going to make decisions later. Whether the writer of such phrases consciously realizes it, he is probably using this device to avoid thinking about the problem or coming to any decisions himself. This phrase will help him to "be right" no matter what happens later. Whether it will help our own decision maker to "be right" is another matter, since the effect of this soothing language will probably be to reassure him that he has lots of time still and there is nothing to be alarmed about yet. He may even infer that when the enemy "makes his decision," the intelligence system will know it and tell him

All non-routine or unusual actions emanating from the national level result from some kind of decisions. They don't just happen. This is true of both military and political actions. In highly monolithic or centrally controlled states (which includes Communist nations), this is even more true than in democracies; i.e. a US governor may call out the National Guard or make some extravagant statement on national policy without its reflecting any decision in Washington, but in the Soviet Union things don't happen this way. Nor, even at the height of the Cultural Revolution in China, would it have been accurate to assume that some independent warlord would have been able to mobilize forces opposite the Taiwan Strait or take some other such action without national authority.

When something unusual occurs, particularly something which increases the adversary's capability to take military action or is otherwise potentially ominous, the analyst should ask such questions as: What does this suggest of enemy plans? What prompted him to do this? What kind of decision has been taken which would account for this action? He should avoid suggesting that the enemy does not know why he did it or that we are waiting for him to make his decision. It will often be helpful at this point to try and look backward and see what may have gone before which could account for the current development or which may indicate that there is a connection between a number of developments. And this in turn may help us in—

b. Isolating or estimating decision times. Major national decisions, and sometimes even minor ones, are likely to result in actions in various fields, all of which flow from the same source or cause and are thus related to one another. They are designed to achieve the same ends or to be complementary. Where the decision is concerned with hostilities or preparations for possible hostilities, it will nearly always be followed by a series of both military and political actions which differ markedly from the norm. In some cases, it will require only a minor amount of backtracking or retrospective analysis to perceive that the actions were probably initiated at a recent publicized meeting of the national leadership, Politburo, Warsaw Pact leaders, or whatever. This will be particularly true if there is some sudden, unexpected development which precipitates a crisis, and ensuing developments clearly follow from that event.

No one should have much trouble in these circumstances in perceiving that decisions of some sort are being taken and when.

Where there is no sudden and obvious emergency, however, both the nature and timing of major decisions are often concealed in closed societies, and sometimes in free societies as well. Thus, it may be some time before there will be indications that any new decisions have been taken at all, let alone when they were taken, or what they might have been. The analyst may often have to work from very fragmentary data in his effort to reconstruct what has been happening up to now and to attempt to determine when the adversary decided to initiate the action. Why bother?

The reason to bother is that the recapitulation of the events or developments in time sequence from the date when the first anomalies became apparent will not only help to fix the decision time but also the nature of the decision. The interrelationship of events as part of a plan may begin to become apparent; they may cease to be isolated, unexplained anomalies when they can be traced back to a common date. Thus we may begin to perceive a scenario in which, for example, the following things began to happen at approximately the same time: reservists were secretly called up; the propaganda line for domestic consumption began stressing the need for greater vigilance against foreign spies; certain key officials were quietly called home from abroad for consultations; a previously scheduled military exercise failed to take place; the leaders of some allied countries went on "vacations" to unannounced destinations; meat became unavailable to civilians in some provinces; large-scale exercises were announced in a border area; a shortage of boxcars began to develop for normal economic needs; a prominent military leader disappeared from public view; the ambassadors of the nation became markedly more friendly in many countries; the number of submarines on patrol began to rise; and so forth. It must be stressed that, in real life, this information will likely be reported in fragments over a period of weeks, never all at once, and some of the developments at the time will seem to have been so insignificant as not to be worth noting, let alone reporting in intelligence publications. Only as they are assembled by the date they were first observed to have occurred (not date they were reported) will the analyst begin to perceive their possible relationships and suspect that some common prior decision may lie behind all or many of them.

Once again, the value of keeping chronologies of bits of seeming incidental intelligence is evident. It is only by doing this that the probable times of secret decisions are likely to be suspected at all, or that the analyst can begin to fit the pieces together. Once it becomes apparent (as it probably will only after meticulous research) that a shift in the propaganda line actually coincided with the first secret mobilization of reservists and a variety of other preparations for possible conflict will the possible scope and significance of the enemy's decisions begin to emerge.

c. Judging that crucial decisions are being made. One of the most important things to know about what the enemy is up to is whether he is making major new decisions at all. That is, even if we have no evidence as yet as to the nature of the decisions, we may gain considerable understanding of the intentions of the adversary if we have some insight into what he is concerned about and whether some particular subject is of overwhelming priority to him at the moment. This is often not so difficult to ascertain as it might appear, although clearly it will be dependent either on what the adversary chooses to publicize about his concerns, or on our ability to collect some information on political developments and the activities and attitudes of the leadership.

Contrary to what many may think, the preoccupation of the leadership with particular problems and decisions may often be no secret at all. To pick a conspicuous example, noted earlier in Chapter 20 and discussed in more detail in this chapter, it was abundantly evident in the summer of 1968 that the Soviet leadership was obsessed with the problem of what to do about Czechoslovakia, and that it had overriding importance to them. It was evident that the Soviet leadership was making decisions of some kind about Czechoslovakia, even if analysts could not agree what those decisions were. The perception of the crisis thus derived in part from our knowledge that Czechoslovakia, from the Soviet standpoint, was what the whole summer was about.

In contrast, we may note that the Sino-Soviet border crisis of the spring–summer of 1969 was not accompanied by similar evidence or suggestions that the Soviet leadership was engaged in decisions of such a crucial and immediate nature concerning what to do about the China problem. This difference should probably not be stressed too much, given the Soviet capability for concealment and deception in this field. Nonetheless, our perception that the crisis in 1969 was not of the same magnitude as that of 1968 derived in large measure from the sense that the Soviet leaders were not taking the same kind of critical decisions. This perception of the degree of criticality of the problem was the result of a variety of information, some of it of a negative nature, that is, we were just not getting the same volume and type of reporting reflecting critical positive decisions that was received in 1968. If this analytic approach sounds highly subjective, it is; but it is of stuff like this that warning judgments are made.

d. Contingency, intermediate and final decisions. All analysts should beware the pitfall of oversimplification of the decision-making process, which is one of the most common of errors. Crucial national decisions usually involve a series of steps, which may range from preliminary decisions to take certain measures on a contingency basis, subsequent decisions to take further preparations and to "up the ante" in case military action becomes necessary, up to near-final and final decisions to proceed with the action. Or, alternatively, actions may be initiated only for pressure purposes or in an attempt to dissuade by threat of force, with no intention of following through.

It may be noted that a nation may make a final or near-final decision as to an objective which it firmly intends to obtain, but will make a series of decisions on the various means which may be tried to obtain that objective. Often, these will involve both political and military pressure tactics, since presumably all-out force is the means to be employed only if all other measures have failed. In this case, a nation may seem to be indecisive (because a series of measures is tried) when in fact it has the objective clearly in mind and always intends to reach it.

In recent years, we have heard much about options, and to "keep his options open" has become a popular phrase to describe various preliminary steps or contingency preparations which the nation may take, presumably when it has not yet decided which course of action it will finally adopt. Indeed the phrase strongly implies "decision deferred," and the more options a nation has, the better off it presumably is and the longer it can defer the crucial decisions.

It is well to avoid over-dependence on this idea, which can lead the analyst into cliche-type reasoning where all preparations for military action, no matter how ominous, are written off as inconclusive and indicative of no decisions on the part of the enemy. The important questions are: "What options are left?" and "Does this action indicate that the adversary himself now believes that the options are dwindling and that the chances for a political solution are running out?" It is surprising how many people seem unable, or unwilling, to carry out this type of analysis, and will fall back time and again on the argument that it was impossible to come to any judgment of the enemy's intentions or decisions since he was "only keeping his options open."

The reader may wish to refer back to the five questions suggested in Chapter 25 as basic guidelines to the interpretation of a nation's course of action. They are also the crucial factors in the decision-making process. If the first question, or premise, is judged to be positive—that the national leadership is committed to the achievement of the objective in question—then this is the operative factor behind the decision, or series of decisions. Only if some other factor effectively prevents or precludes obtaining that objective, will the nation presumably be deterred from a course of action which will fulfill its objective. Some of the means it may use (which we describe as options) may indeed be contingency or preparatory steps initially, in case other more desirable options fail or do not prove viable, but they are means to an end, not just steps taken to have "more options" and hence to postpone coming to any decisions. Indeed, the number of options which the nation devises, or tries out, to secure its objective may be something of a rough measure of how serious it is about obtaining it. As applied to Czechoslovakia in 1968, the mere fact that the USSR tried so many means of bringing the situation under control before it invaded was in itself indicative of the seriousness of its intent and raised the probability that the military option ultimately would be exercised if all else failed.

e. Inter-play of political and military decisions. Another simplistic approach to the decision-making question, which also occurs surprisingly frequently, is to assume that political and military decisions are taken by different groups and are somehow not realty related to one another. On the one hand are political leaders making political decisions and on the other are military leaders undertaking military exercises, carrying out, mobilization, deploying troops, etc., almost on their own without relationship to the political situation. This is highly erroneous, at least in countries where the national leadership exerts effective command and control over the military forces, and it is particularly erroneous in Communist nations in which the political leadership maintains a monopoly on the decision-making process and the military undertakes virtually nothing on its own. The Party runs the Soviet Union, and the Politburo makes the decisions.

Thus, military and political decisions are inter-related and part of the same process, and the military steps are undertaken and in so far as possible timed to achieve specific political objectives. They must not be considered in isolation or as unrelated to the political objective. To do so is not only to misunderstand the cause of the military actions, but more importantly to fail to perceive the strategic objective and the interrelationship of the various means which may be used to obtain it.

Soviet Decision-Making in the Cuban Missile Crisis

Among the many unresolved mysteries of the Cuban missile crisis are why the USSR undertook the action at all, how and when the Soviet decisions to undertake it were made, and what the outcome of this highly dangerous adventure was expected to be. I have often asked analysts who profess to understand this crisis what they think Moscow was going to do *after* it succeeded in establishing an operational missile force 90 miles from the continental US. The fact is that no one knows, and we can only speculate. The Soviet plan may be compared to a drama in which we missed the first act altogether, came in for the second act and forced the author to bring down the curtain because we did not like the plot, and never did learn what denouement was planned for the third act.

Nor are we likely ever to know. The most exhaustive research after the event and numerous retrospective analyses have failed to shed much light on the crucial questions of Soviet motivations, planning, decisions, and expectations. The post mortems have taught us more about *what* happened that summer and fall (although even here there are still many gaps), but virtually nothing more about *how* it was decided and *why*. (It may be added parenthetically that Khrushchev's so-called memoirs, whether or not he wrote them, are almost no help and very misleading.)

It is often possible, after a crisis is over, to come to some fair understanding of when certain decisions were taken, even though it was not apparent at the time. In the case of the Cuban missile crisis, we cannot even establish this. There is no evidence when or even that high-level discussions were held with the Cubans, since there is no record of an exchange of VIP visits during the period when the crucial decisions must have been made. The only thing that appears reasonably clear is that the Soviet decision would have been made some time between the autumn of 1961 (when the USSR's efforts to obtain a Berlin settlement had failed) and the early spring of 1962. Based on the magnitude of the subsequent logistic effort and the extensive planning that must have gone into coordinating the various shipments of materiel and troops to Cuba, it has generally been considered that the decision was probably made no later than April, or some three months before the first ship movements were detected in July. During the entire period from the fall of 1961 to July, we had almost no indication that the USSR was particularly preoccupied with Cuba, or that any important decisions with regard to that island were being made. The seeming focus of Soviet concern and the area of greatest danger of confrontation continued to be Berlin. Thus Soviet security effectively concealed that *any* highly dangerous and crucial decisions were being taken concerning a confrontation with the US and also that anything unusual was under way with regard to Cuba.[1]

It may also be observed that, even in retrospect, we do not know who thought up the plan (although we presume it was Khrushchev himself, it could be that he adopted the ideas of someone else), which members of the Soviet leadership favored or disapproved the idea, or what the attitude of the military was. It can be argued, with almost equal plausibility, that Soviet military leaders would have opposed so risky a venture which could have led to hostilities, or that they would have favored a measure designed to alter the strategic balance of power prior to undertaking certain other dangerous ventures, for example, in Berlin.

Intelligence, both at the time and in retrospect, also has shed very little light on the Soviet decision-making process from the time that the US detected the shipments to Cuba and began to become concerned about them up to the discovery of the strategic missiles in mid-October. It is quite evident that the various US warnings to the USSR of the dangers of introducing offensive weapons into Cuba did not serve to deter Khrushchev and presumably also had no effect on the nature or timing of the shipments to Cuba. We do not know, however, whether these warnings prompted new debates in the Kremlin on the whole question and whether they were partially responsible for some of the military readiness measures which the USSR initiated about in September (the day it announced that its forces were being brought to "highest combat readiness").

President Kennedy's speech of 22 October announcing the discovery of the strategic missiles and the imposition of the US "quarantine" of course precipitated a week of overt crisis in which it was quite evident that the USSR was taking some very important decisions indeed. The first of these which became evident to us was the order to halt the movement toward Cuba of ships carrying military equipment, a highly important indication that the USSR desired to avoid a direct confrontation with US forces, at least at sea. In subsequent days, however, work continued on the missile sites in Cuba. As everyone knows, the final and crucial evidence of the Soviet decision to back down was conveyed in the messages sent to President Kennedy by Khrushchev, first privately on 26 October and again, when it was made publicly, on 28 October, in which the USSR agreed to withdraw the missiles in return for US assurances that there would be no invasion of Cuba. On the intervening day (27 October), there appeared to be real indecision in the Kremlin, since on that day Moscow radio carried a statement by Khrushchev which said that the USSR would remove its missiles from Cuba if the US did the same from Turkey.

There are a number of indications that Khrushchev became convinced on 27 October that the US had decided to invade Cuba and that the attack was imminent. This was the date when Soviet forces apparently were brought to their highest readiness, and the crisis was further heightened by the shootdown of a U-2 over Cuba (an act, incidentally, which almost certainly was not ordered by Moscow). Khrushchev, speaking to the Supreme Soviet on 12 December 1962, stated that, "In the morning of 27 October, we received information. which directly stated that this attack [on Cuba] would be carried out within the next two or three days." There appears little doubt that it was this threat of imminent attack on Cuba, or actually on Soviet missile installations and forces in Cuba, which led to the final and publicly announced Soviet decision to withdraw the missiles. We also have little reason to doubt that the decision was Khrushchev's own, although again we have no insight into what role the rest of the leadership may have played in that decision. There were subsequent hints in the Soviet press that elements of the military were dissatisfied with the decision, but in the USSR the military is not represented in the highest decision-making body or Politburo.

As a footnote to the inter-play of decisions by both sides, it is interesting to note that Robert Kennedy's book on the missile crisis, published several years later, revealed that President Kennedy had not reached a final decision on 27 October to proceed with an attack on Cuba. "We won't attack tomorrow, the President said. We shall try again."[2] Nonetheless, the President was apparently ready to order the attack if necessary. That evening, Robert Kennedy informed Ambassador Dobrynin that the US must have a commitment by the next day that the missiles would be removed, or otherwise the US would remove them.[3] Further, there were leaks to the press quoting US leaders as indicating that such action would be taken soon, and these, together with the major US military steps, appear to have been convincing to the USSR.

The Cuban missile crisis, which to many analysts is both the most fascinating and most elusive problem we have ever had, is an example of an almost total failure to have perceived the nature of the adversary's strategic decisions, up to the moment when conclusive evidence was finally obtained. Not only would it have been impossible, on the basis of the information available at the time, to have reconstructed the decision-making process; it is almost impossible to do so even in retrospect. We must rely on hindsight even to perceive that decisions were being taken, let alone what they were.

If the Cuba crisis represents the nadir of our collection and perception of an adversary's decision-making process, the Czechoslovak crisis six years later provides an altogether different problem, one in which we had ample evidence that Soviet decisions were being taken on Czechoslovakia, but in which there is still considerable controversy and misunderstanding concerning what decisions were taken and when. Thus, the Czechoslovak instance provides an excellent example of the process of reconstructing decision-making for warning purposes, which would never have been possible in the Cuban crisis.

Soviet Decision-Making in the Czechoslovak Crisis

From the time that Novotny was replaced as head of the Czechoslovak Communist Party in January up to the Soviet-Warsaw Pact invasion of Czechoslovakia on the night of 20–21 August 1968, there was abundant and continually mounting evidence of Soviet concern with the problem of Czechoslovakia and what could be done to reverse the course of liberalization undertaken by the Dubcek regime. The controversy, and there was much controversy, over Soviet intentions never arose from lack of evidence of Soviet preoccupation with Czechoslovakia. Since the Soviets made no secret of their concern, and since most of the various meetings with their allies (the so-called "Warsaw Five") were public knowledge, as was much of the pressure which they brought against Czechoslovakia, analysis of their decision-making becomes one of interpreting from the available evidence what the decisions were, not whether decisions were being taken at all.

Chronology of Major Developments Prior to the Invasion of Czechoslovakia

23 Mar: Trends in Czechoslovakia criticized at Dresden conference, attended by leaders of Czechoslovakia, the USSR, East Germany, Poland, Hungary and Bulgaria.

9–10 Apr: Soviet Central Committee Plenum heard unpublished speech by Brezhnev, reportedly dealing particularly with Czechoslovakia, after which Soviet leadership toured the country to explain situation to local party meetings.

24–25 Apr: Marshal Yakubovskiy, Commander-in-Chief of Warsaw Pact, visited Prague and reportedly asked to have Soviet or other Pact forces stationed in Czechoslovakia. Request refused.

4–8 May: Czechoslovak leaders in Moscow for unsatisfactory talks, followed by separate and unexpected visit of leaders of East Germany, Poland, Hungary and Bulgaria.

6–12 May: Several Soviet divisions deployed to positions near Czech border in East Germany, Poland and USSR.

17–24 May: Soviet Defense Minister Grechko and other Soviet military leaders in Prague, with concurrent visit by Premier Kosygin; followed by announcement that Warsaw Pact exercises would take place in Czechoslovakia and Poland in June.

1–23 Jun: Soviet troops which had deployed to borders moved into Czechoslovakia in undetermined strength, but apparently substantially more than token forces publicly announced.

20–30 Jun: Warsaw Pact exercise "Sumava" conducted in Czechoslovakia, after which Prague and Moscow announced conclusion of exercises, but Moscow retracted its announcement.

1–10 Jul: Soviet forces lingered in Czechoslovakia, amid mounting signs they might not leave.

10–15 Jul: Czechoslovak leaders refused to go to Warsaw for talks. The other five met there without them and issued very tough letter calling on Prague to take steps to reverse trend toward liberalization. Prague announced Soviet troop withdrawal begun.

17–19 Jul: CPSU Central Committee Plenum convened suddenly to endorse actions taken at Warsaw. Moscow again demanded talks with Czech leadership. Marshal Grechko returned to Moscow from Algeria ahead of schedule.

20–23 Jul: USSR spreading word it might have to use force in Czechoslovakia. Letters dispatched to friendly Communist Parties warning that USSR would use any necessary means, including force, to bring situation under control. Ranking Polish Party member reportedly said decision made by USSR to break Czech will by force if necessary.

23–30 Jul: USSR announced that large-scale Rear Services exercises would be held over the entire western USSR lasting until 10 Aug and would involve callup of reservists, requisitioning of transport and demothballing of equipment. Extension of these exercises into East Germany and Poland, and involvement of forces of those countries, announced on 30 July.

24–31 Jul: USSR moved major forces to the border of Czechoslovakia, bringing total divisions to an estimated 15–19, with major movements being observed in East Germany and Hungary. Ground force movements accompanied by major deployments of tactical air units. There were indications that some East German and Polish units also deploying to border.

28–30 Jul: Western travelers sighted large movements of Soviet troops and supply columns into Poland at several crossing points from the USSR, ranging from southern Poland to the Baltic coast. Large holding areas for Soviet forces in central Poland found by Western attachés in early Aug.

15 Jul–3 Aug: Very slow withdrawal of Soviet forces from Czechoslovakia under way. Attaché sighting on 29–31 July finally confirmed presence of division-size Soviet unit in Czechoslovakia.

29 Jul–3 Aug: Meeting of Soviet and Czech leaders at Cierna and of Warsaw Five leaders with Czechs at Bratislava resulted in ill-defined Czech commitment to take steps to strengthen Party control and to strengthen ties with Warsaw Pact. Public polemics to cease. USSR also reported to have again demanded at Cierna that Soviet troops be stationed in country.

3–16 Aug: Crisis appeared eased, but Soviet-Pact forces remained deployed and military preparations were continuing. Czechoslovakia was failing to take effective measures to carry out agreement and was again warned by Soviet press of need to comply.

11–16 Aug: USSR announced that combined communications exercise being conducted in USSR, Poland, and East Germany, and that top military commanders meeting in East Germany and Poland.

16–20 Aug: Soviet press attacks on Czechoslovakia increased, but contained no explicit warning of imminent military action. Press reports, 20 Aug, indicated Soviet leaders again meeting in Moscow.

20–21 Aug: Soviet and Warsaw Pact forces conducted massive invasion of Czechoslovakia.

To simplify the subsequent discussion, the preceding two pages recapitulate in capsule form a chronology of the major events leading up to the invasion of Czechoslovakia, beginning with March. It is emphasized that this presents only highlights and should not be considered anything approaching a complete indications roundup.

Working from the facts presented in the chronology (which has excluded any information obtained only after the invasion), when were the major Soviet decisions taken and when did the USSR "decide" to invade Czechoslovakia? The student of this problem should be advised that there has never been agreement on this subject among analysts. Post mortems revealed the widest variations in opinion, even among those fairly knowledgeable on the facts, and those who did know many details of the available evidence not surprisingly were still hazier on what might have happened; and when.

Most of the argument concerning Soviet decisions has centered around the period of mid-July and the period 16–20 August, and in essence the question is: Did the USSR "decide" to invade at the time of the Warsaw meeting with its allies, or did it only initiate the contingency military preparations at that time, thus "keeping its options open" for military invasion if it should later decide to take that course of action? Were the massive military preparations which became evident to us in the last week of July taken "just in case" or even taken by the military more or less on their own, or do they indicate a prior political decision that force would be required to bring the situation under control? Anyone who has not participated in debates on this subject might be amazed at the extent and even bitterness of controversy on this point, not only at the time but long after the event. Some proponents of the view that the USSR did not "decide" to invade until about 17 August have even seemed to suggest that the presence of all those military forces in the field for "exercises" was almost fortuitous, and that the leadership belatedly recognized Marshal Grechko's foresight in deploying those troops and decided suddenly to make use of them to invade. It has been suggested that the leaders were making the invasion plans, or at least working out the details, as late as two or three days before it occurred.

A less extreme, and much more prevalent, view concedes that this is ridiculous and that the military operations obviously had been carefully planned well in advance, but nonetheless holds that it was "impossible" to make a judgment of Soviet intentions from the evidence at hand prior to the invasion and that the USSR itself did not know what it was going to do until three or four days before. This view was reflected in conclusions put forth in intelligence publications for three weeks prior to the invasion—that Soviet forces were in a high state of readiness to invade "if it was deemed necessary" or if the USSR "decided to do so," thus conveying to the reader the sense that not only we but the Soviets themselves did not yet know what might occur, and that the crucial decisions were yet to be taken. Thus the intelligence judgment went as far as it could, and certainly was not wrong, even though the Soviets did finally "decide" to invade.

This controversy over the timing and nature of Soviet decisions well illustrates many of the points set forth in the opening portion of this chapter. Most comment, and thinking, on the Soviet decision-making process on Czechoslovakia has been highly simplistic, sometimes even naive. For the most part, analysts have failed to examine the known developments in sequence in an attempt to understand what decisions should logically have been taken *before* something happened, or to ask whether a series of observable actions are related to one another and may be logically traced to a common prior decision. Analysis has been overly concerned with what happened rather than with what caused it to happen. In the process—and this is a most serious error—it has often failed to distinguish between the Soviet *objective* and the *means* employed to achieve that objective, and thus has failed to perceive that fundamental and important decisions were probably taken much earlier than is usually believed.

Seldom in these discussions has it been suggested that major Soviet decisions might have been made earlier than mid-July, or that it is pertinent to examine what the USSR was trying to accomplish before that time. Many observers have maintained that Soviet policy was vacillating or inconsistent and that the leadership seemingly was unable to make up its mind what to do about Czechoslovakia, right up until the time (unspecified) when the Politburo finally voted to invade. There are a number of reports, authenticity largely unknown and frequently contradictory, which purport to set forth how the Politburo voted on the invasion and which members were for or against the decision.

To which the perceptive analyst of the decision-making process must ask: What vote are you talking about? The Politburo did not vote just once, or make just one decision, on Czechoslovakia. It is clear from the chronology of the evidence that there must have been a series of decisions—and, moreover, that these decisions were not logically inconsistent with one another and did not reflect vacillation on the part of the Soviet leadership. By this we do not mean to say that there was no controversy over what to do, or that the final outcome followed inevitably from the first crucial decision, but only that the scenario does have a logical plot, and that the timing and probable nature of the decisions can be reconstructed with considerable accuracy from the events which ensued.

The Soviet *objective* was the restoration of orthodox Party control in Czechoslovakia which would ensure that the country would remain a faithful Soviet political ally and a full military participant in the Warsaw Pact. To the USSR, given Czechoslovakia's geographic location, such control was essential to the Soviet security position in Eastern Europe, and the defection of Czechoslovakia could have had incalculable in adjoining countries as well. As the "Prague spring" flourished the liberalization program increasingly permitted a freer atmosphere which the USSR perceived as a threat to the alliance, the Soviets were concerned with finding a *means* of restoring orthodox Party control. Politically, this might have been accomplished in one of two ways: by persuasion or pressure on Dubcek to see the light and himself undertake the necessary steps to reverse the trend toward liberalization; or by the replacement of Dubcek and the "liberals" with an orthodox conservative leadership. In fact, the USSR tried both these means, the "persuasion" and pressure continuing virtually up to the invasion. Although little is known of Soviet attempts to overthrow Dubcek by a coup within the Party, there is reason to believe that the USSR hoped, at least initially, and perhaps even tried, to do this, but that the conservative element was entirely too weak to accomplish it without more support than the USSR could bring to bear.

Another means which could be used to help secure the political objective was military power. It is important to understand that force, either present or readily available, is an essential ingredient of dictatorial control, that the Communist regimes in Eastern Europe have remained in power only because of the proximity of Soviet forces, and that this is a major reason why Soviet troops are deployed in Eastern Europe. The normal Soviet response to any serious threat of political dissidence is to deploy more military force to the area. Thus, in Moscow's eyes, the introduction of Soviet forces into Czechoslovakia was a means by which its political objectives would be furthered. The objective was not to carry out a massive invasion of Czechoslovakia—this was undertaken only when all other measures had failed to achieve the desired result—but rather to get Soviet forces into the country by one means or another, so that their presence would serve to keep the lid on the political situation. The evidence indicates that the USSR was attempting from April onward to introduce Soviet forces into Czechoslovakia—not because it sought a "military solution" but because the troops were to be a means of obtaining the desired political solution. In this sense, the USSR intended from

April onward to "invade" and in fact it conducted not one, but two, invasions of Czechoslovakia.

Let us now go back to what happened that spring and summer, insofar as we have outlined it in the previous chronology, and try to reconstruct the Soviet decision-making process. In all, we can isolate five different occasions in fact almost the precise dates—on which the Soviet leadership, either alone or in concert with its allies, must have decided on some action concerning Czechoslovakia. Further, we can from the events which followed come to logical and probably quite accurate judgments as to what the essence of the decisions must have been. Table 27.1 summarizes this information on the decisions, including a little information on the period from 16 August onward which did not become available to us until after the invasion.

If this reconstruction of events is essentially accurate, as I believe it to be, it will be seen that the basic political decision—to restore effective control in Czechoslovakia and to keep it within the Warsaw Pact—had probably been taken in principle by early April. This does not, of course, necessarily mean that all subsequent measures undertaken were then discussed, or that a final requirement for full-scale invasion was then foreseen, although the military almost certainly would have been directed to begin contingency planning for this possibility, once the political decision had been reached. Within a month, deployments were undertaken with the intention of "invading" under cover of a Warsaw Pact exercise, the scope and purpose of which had been grossly misrepresented to the Czechoslovak leaders. Whatever doubts we may have had about the purpose of all this were, or at least should have been, dispelled when the Soviet troops failed to leave the country after the exercise despite mounting Czechoslovak concern and demands. Finally, controversy over, the nature and size of the Soviet forces which had entered for the "exercise" was largely resolved by attaché sightings the end of July which confirmed the presence of at least one full division, fully combat prepared, and not the usual token forces used in Warsaw Pact exercises.

So what did the Pact allies discuss in Warsaw in mid-July, following which the mobilization and deployments of the invasion forces occurred? What is meant by those who maintain that no decisions were then taken on future actions, but only "contingency" preparations in case it was later decided to use force, and that the Soviets themselves never made up their minds until three or four days before the invasion? I personally find this reasoning extremely shallow and non-perceptive, and illustrative of a fundamental error in the warning process—to wit, a reluctance to come to judgments on the grounds that the enemy has yet to make his decision, so how can we tell what he will do? I will try to explain why this type of reasoning does not hold water, at least in this instance, recognizing, however, that this will not be convincing to all

DATE	EVENT	PROBABLE NATURE OF DECISIONS	RESULTING ACTIONS
9-10 Apr	Soviet Central Committee Plenum	Endorsed Politburo decision that control of Czechoslovakia must be maintained and that Soviet troops should be placed in country.	Party membership informed. Yakubovskiy dispatched to ask, unsuccessfully, that Soviet troops be stationed in Czechoslovakia.
4-8 May	Meetings in Moscow of Soviet and Czech leaders, and of Soviets with other four countries	Four Eastern European countries and USSR agreed on further steps, including deployments of Soviet troops to Czech border, and possibly to try to oust Dubcek by internal Party coup.	Soviet troops deployed to Czech border. Czechs persuaded to permit Soviet troops to enter country for Warsaw Pact "exercise." Deployed forces conducted "silent invasion" of Czechoslovakia.
14-15 Jul	Meeting of Warsaw Five, following Czech refusal to talk with Soviets	Agreement of the Five that political solution unlikely, and that full-scale invasion would probably be required to force Czech compliance. All military preparations for invasion to be carried out, but another effort also to be made at political solution.	Mobilization and deployment of Soviet and Warsaw Pact forces. Heavy pressure brought on Czechs to meet. Soviet Central Committee Plenum endorsed any action leadership considered necessary, giving it free hand thereafter. Friendly parties informed force would be used if needed.
29 Jul - 3 Aug	Cierna and Bratislava conferences	Soviet leaders probably decided and their allias agreed to give Czechoslovakia one more chance to comply. Final decision to invade probably deferred.	USSR toned down polemics against Czechs but continued military preparations for invasion, announcing a series of "exercises" as cover. Czechs failed to tighten controls as promised.
16-17 Aug	Meetings of Soviet leaders (but not Central Committee) in Moscow	Took final decision to proceed with invasion on night of 20-21 Aug; considered last-minute measures -- including agreement to begin SALT with the US and invitation to President Johnson to visit USSR.	Final military preparations for the invasion completed. Propaganda against Dubcek increased but without direct warning of imminent invasion. US informed of Soviet agreement to begin SALT talks.

Table 27.1.

readers, perhaps not even to most. But it is the heart of the warning problem, and the difference between those who "had warning" of the invasion of those who did not.

First, the course of events up to mid-July, as well as a great deal of information received in the following two to three weeks, provided a convincing amount of evidence that the Soviet leadership was desperately seeking to bring the Czechoslovak situation under control, and that nothing held higher priority to them than this. There was a true crisis atmosphere, which belied any judgment that the Soviets were exaggerating their concern or bluffing. Moreover, we received virtually unequivocal warning by the end of July that foreign non-ruling Communist Parties had been informed that the USSR would take whatever means were necessary, including force, to bring the situation under control and, even more explicitly, a reliably sourced report that the decision *had been taken* to break the Czechoslovak will by force if necessary.

The scale of the military buildup also was consistent only with a decision to carry out preparations for full-scale invasion, and could not reasonably be dismissed as "exercises," an attempt to bring "more pressure," or preliminary contingency measures. This fact was recognized and reported, at least to the extent that the forces were held to be in a "high degree of readiness" to invade. Moreover, in addition to the forces deployed along the Czechoslovak border, the movement of substantial additional Soviet forces into northern Poland (apparently as backup or reserve forces) provided further evidence of a bona fide buildup of combat forces rather than mere "pressure" on the borders of Czechoslovakia.

What, then, was the most probable nature of the decisions taken in mid-July, and again a few days before the invasion, and which was the most important? Which was "the decision, or "the" Politburo vote sometimes referred to as if there had been only one? It appears to me that the crucial decisions with respect to the invasion must have been taken in mid-July, and that the massive military buildup was initiated *because* the USSR and its allies became convinced that a solution by political means was unlikely, and that a military invasion would therefore probably be required to bring the situation under control Or, to put it another way, the Soviet leadership reached the basic decision in mid-July to carry out an invasion unless a political solution could be reached. At the same time, it was apparently also decided to make one more effort toward a political solution before proceeding with invasion, and thus probably to defer a final decision on whether military action would inevitably be required—and hence also the decision on its timing.

And what was "decided" on 16–17 August that required the meetings of the leadership (but not, we may note, the reconvening of the full Central Committee which had already given its endorsement of the mid-July decisions)?

So far as we can tell, these meetings were concerned with two things: a final decision to proceed with the invasion on the night of 20–21 August, and the last-minute details of the plans—including the message to President Johnson agreeing on SALT talks. We remain uncertain why this date was chosen, although it s likely that both political considerations (the final acceptance that Dubcek was either unwilling or unable, or both, to implement the Cierna and Bratislava agreements in the sense in which the USSR intended) and military considerations (the forces were fully ready and probably had been so for at least several days) prompted the final decision to go ahead. Although this decision was not, of course, "inevitable"—the basic decisions of mid-July could always have been reversed and the invasion called off—the most crucial decisions and votes were nonetheless those of July. From then on, invasion was always a probability barring some political miracle, which the Soviet leadership really did not expect. Moreover, the action followed logically from the earlier decisions, almost certainly dating back to April, that by one means or another Czechoslovakia must be kept within the Warsaw Pact.

The analysis above, while convincing to me, will not necessarily be so to others, particularly those who have long maintained that the USSR did not "decide" to invade until 16–17 August. We cannot answer the question definitely; the evidence does not permit an absolutely firm conclusion and will continue to be interpreted differently by different individuals. The case does illustrate the difference in reasoning, however, between the individual who says that the Soviets were "just keeping their options open" and then stops analyzing, and the individual who asks "what options did they have left?" To understand this difference is truly to perceive the warning problem on Czechoslovakia.

Notes

1. It is true that there were a very few and seemingly unrelated clues from clandestine sources which indicated this possibility, but it would have required almost clairvoyant perception to have anticipated the Soviet move before the shipments to Cuba began—and no one would have believed it. There were some other indications that summer that something potentially big might be brewing in the minds of the Soviets, but its nature and any connection with Cuba could hardly have been foreseen.

2. Robert F. Kennedy, *Thirteen Days, a Memoir of the Cuban Missile Crisis* (New York, W. W. Norton & Co., 1969), p. 101.

3. pp. 108–109.

28

Assessing the Timing of Attack

ONE OF THE MOST WIDESPREAD MISCONCEPTIONS about warning is the belief that, as the hour of the enemy attack draws near, there will be more and better evidence that enemy action is both probable and imminent. From this, the idea follows naturally that intelligence will be better able to provide warning in the short term and will, in the few hours or at most days prior to the attack, issue its most definitive and positive warning judgments. Moreover—since there is presumed to be accumulating evidence that the enemy is engaged in his last-minute preparations for the attack —this concept holds that intelligence will likely be able to estimate the approximate if not the exact time of the attack. Therefore, if we can judge at all that the attack is probable, we can also tell when it is coming.

This concept of warning—as a judgment of imminence of attack—has adversely affected US thinking on the subject for years. As of this writing, the official definition of strategic warning in the *JCS Dictionary* is, "A notification that enemy-initiated hostilities may be imminent." (See discussion of this question in Chapter 2). More explicitly, the US national warning estimate of 1966 concluded: "Intelligence is not likely to give warning of *probable* Soviet intent to attack until a few hours before the attack, if at all, Warning of increased Soviet readiness, implying a *possible* intent to attack, might be given somewhat earlier."

However logical these suppositions may appear in theory, they are not supported either by the history of warfare nor the experience of warning analysts, and in recent years more realistic assessments of this problem have begun to appear in warning papers and estimates.

For the fact is that warning judgments are not necessarily more accurate or positive in the short term and that assessing the timing of attack is often the most elusive, difficult and uncertain problem which we have to face. It is simply not true that the last few days or hours prior to the initiation of hostilities are likely to bring more and more specific indications of impending attack which will permit a better or more confident judgment that attack is likely or imminent. In many cases experience shows that the reverse will be true, and that there will be fewer indications that the attack is coming and even an apparent lull in enemy preparations. This can be quite deceptive, even for those who know from experience not to relax their vigilance in such circumstances. Those who do not understand this principle are likely to be totally surprised by the timing—or even the occurrence—of the enemy action. They will probably feel aggrieved that their collection has failed them and they will tend to believe that the remedy for the intelligence "failure" is to speed up the collection and reporting process, not appreciating that the earlier collection and analysis were more important and that a judgment of probability of attack could have been reached much earlier and should not have been dependent on highly uncertain and last-minute collection breakthroughs.

Principal Factors in the Timing of Attacks and the Attainment of Surprise

Nearly all nations, except in unfavorable or unusual circumstances, have shown themselves able to achieve tactical surprise in warfare. History is replete with instance in which the adversary was caught unawares by the timing, strength or location of the attack—even when the attack itself had been expected or considered a likelihood. Even democracies, with their notoriously inept security in comparison with closed societies, have often had striking success in concealing the details (including the timing) of their operations. To cite the most conspicuous example, the greatest military operation in history achieved tactical surprise even though it was fully expected by an enemy who potentially had hours of tactical warning that the massive invasion force was approaching. It was the Normandy invasion. In Chapter 29 we will discuss the role which deception played in this operation.

It is not only by deception, however, that tactical surprise is so often achieved and that last-minute preparations for the attack can be concealed. A more important and more usual reason is that the indications of attack which are most obvious and discernible to us are the major deployments of forces and large-scale logistic preparations which are often begun weeks or even months before the attack itself. Once these are completed, or nearly so, the enemy will have attained a capability for attack more or less at the time of his choosing, and the additional preparations which must be accomplished shortly prior to the attack are much less likely to be discernible to us or may be ambiguous in nature. Staff conferences, inspections, the issuance of basic loads of ammunition and other supplies, and the final orders for the attack all are measures which require little overt activity and are not likely to be detected in time except by extraordinarily fine collection and rapid reporting—such as a well-placed agent in the enemy's headquarters with access to some rapid means of communications, or the fortuitous arrival of a knowledgeable defector. Even the final deployments of major ground force units to jumpoff positions for the assault may be successfully concealed by the measures which most nations take to ensure tactical surprise—including rigid communications security and night movements. Thus, unlike the major deployments of troops and equipment which almost never can be entirely concealed, the short-term preparations have a good chance of being concealed, and quite often are. And, even if detected, there will often be minimal time in which to alert or redeploy forces for the now imminent attack, still less to issue warning judgments at the national level. Such tactical warning usually is an operational problem for the commander in the field. Ten minutes or even three hours warning does not allow much time for the political leadership to come to new decisions and implement them.

Another facet of the problem of assessing the timing of attack is the difficulty of determining when the enemy's preparations are in fact completed, and when he himself will judge that his military forces are ready. As we have noted elsewhere, it is particularly difficult to make this judgment with regard to logistic preparations. In fact, I can recall no instance in my experience in which it could be clearly determined that the logistic preparations for attack were complete, particularly since heavy supply movements usually continue uninterrupted even after the attack is launched. There has often been a tendency for intelligence to believe that all military preparations are completed earlier than in fact is the case—the discrepancy usually being attributable to the fact that the major and most obvious troop deployments had apparently been completed. Thus, even when intelligence has come to the right judgment on enemy intentions, it has sometimes been too early in its assessment of the possible timing of the attack.

In addition, the enemy command for various reasons may not go through with an attack as soon as the forces are fully prepared, or may change the date of the attack even after it has been set. A recent study has compiled some data concerning the frequency with which D-Days are not met, and the effects of this on the adversary's judgments. Of 162 cases analyzed where D-Days applied, almost half (about 44 percent) were delayed, about five percent went ahead of schedule, and only slightly more than half (about 51 percent) remained on schedule. The most common reasons for delay were weather and administrative problems, presumably in completing or synchronizing all preparations. Some attacks have had to be postponed repeatedly. For example, the Germans' Verdun offensive of 21 February 1916 was postponed no less than nine times by unfavorable weather.[1]

Such changes in plans have sometimes had notable effects on the opponent's assessments, particularly when he has gone through one or more alerts of impending attack which failed to materialize. Whaley notes that the finding that procrastination can help to generate surprise is explainable by the "cry-wolf" syndrome—whereby the false alert, and particularly a series of them, breeds skepticism or downright disbelief of the authentic warning when it is in fact received. "Moreover, the trend is that the greater the number of false alerts, the greater the chance of their being associated with surprise. . . . [The] Aesopian moral seemingly holds . . ., the false alarms serving mainly to undermine the credibility of the source and dull the effect of subsequent warnings. . . . It is ironic that . . . some of the D-Day warnings were quite authentic, the enemy having merely unexpectedly deferred the operation. The consequence was, of course, that several superb intelligence sources including Colonel Oster, Sorge and Rossler received undeserved black marks on the eve of their subsequent definitive alerts.[2]

Of all aspects of operational planning, the easiest to change and most flexible is probably timing. Once troops are in position to go, orders to attack usually need be issued no more than a few hours ahead, and the postponement of even major operations rarely presents great difficulties to the commander. Attacks have been postponed—or advanced—simply because there was reason to believe that the enemy had learned of the scheduled date. Obviously, among the simplest of deception ruses is the planting of false information concerning the date of operations with the enemy's intelligence services.

In addition to general preparedness, tactical factors and surprise, operations may be delayed for doctrinal reasons or to induce enemy forces to extend their lines of communication or to walk into entrapments in which they can be surrounded and annihilated. The delayed counteroffensive, designed to suck enemy forces into untenable advanced positions, is a tactic which the Communists have employed with devastating effect. Obviously, misjudgments of the enemy's intentions in such cases have been heavily influenced by the seeming delay in his response, which induces a false sense of security that he will not respond at all.

Political factors also may weigh heavily or even decisively in the timing of operations. This, of course, will be particularly true when (as is often the case) the nation in question intends to resort to military operations only as a last resort and hopes that the threat of such action will induce the opponent to capitulate. Obviously, in such cases, the decision of the national leadership that the political options have run out and that only force will succeed will be the determining factor in when the military operation is launched (see Chapter 27). In this event, operations may be deferred for weeks beyond the date when military preparations are completed, and the assessment of the timing of the attack may be almost exclusively dependent on knowledge of the political situation and insight into the enemy's decision-making process.

Still another political variant which may affect the timing of attack is when one nation is attempting to induce the other to strike the first major blow and thus appear as the aggressor. In this case, a series of harassments, border violations and various clandestine tactics may be employed as the conflict gradually escalates until one or the other power decides to make an overt attack. Clearly, the point at which this may happen will be very difficult to predict.

Apart from the various reasons noted above, there may be other largely tactical considerations which will affect the timing of attack. Weather, as already mentioned, is one of these—not only visibility, but in some cases winds, tides, moonlight or lack of it. Conditions of roads and terrain of course have been a major determining factor in when some operations will be launched. Military operations and logistic movements of Communist forces in Southeast Asia have traditionally been greatly slowed, if not halted altogether, at the height of the rainy season, and spring thaws on the plains of central Europe have delayed many operations. In cases where weather effectively precludes overland movement, it is of course highly probable that attacks will not occur. Nonetheless, there is always a chance that an enemy may choose to attack even in highly adverse conditions in the interests of achieving surprise.

As is well known, many attacks are initiated near dawn, for two reasons: the nighttime cloaks the final deployments of the attacking units, and the hours of daylight are desirable to pursue the operation. Several Communist nations, however, have shown a marked favoritism for attacks in the dead of night. This has been particularly true of North Vietnamese and Viet Cong forces, which have shown themselves highly adept in night penetration operations and assaults. The USSR also has often launched attacks or other operations hours before dawn: the operation to crush the Hungarian revolt began between about midnight and 0330; the Berlin sector borders were sealed about 0300; the invasion of Czechoslovakia began shortly before midnight.

The USSR also has shown some favoritism for Sunday, both the Hungarian and Berlin operations having occurred in the early hours of a Sunday morning. It would be dangerous, however, to assume that this would be the case. The invasion of Czechoslovakia occurred, for instance, on a Tuesday night, slightly to the surprise of some who had come to expect Soviet operations to begin on Sundays. Whaley has found some preference for Sunday operations among Communist states but not in a majority of cases; it was true in only about one-fourth of the operations which he studied.[3] Among other nations, there does not appear to be any evident preference for particular days of the week. In cases where Sunday is chosen, it is not for any anti-religious reason, but because the alert status of most Western nations is then usually lowest. The Japanese selected Sunday for the Pearl Harbor attack because their observations had shown that most US ships would then normally be in port.

Some Examples of Problems in Assessing Timing

Because of space limitations, discussion of more than a few examples is precluded, and even these must be covered briefly. There is considerable military historical writing, particularly on World War II, which may be consulted by those who wish to study this aspect in more detail, as well as the many examples in Whaley's previously cited work. Since much of this material is readily available and the timing of the Normandy invasion also is addressed in Chapter 29 of this book, the examples below include only two from World War II with the remainder drawn from more recent intelligence experience.

The German attack on Holland, Belgium and France, May 1940. World War II had been under way for eight months before Hitler finally launched his offensive against Western Europe in May 1940, the long delay in the opening of the western front having generated the phrase "phony war." All three victims of the final assault had ample and repeated warnings, and indeed it was the redundancy of warnings which in large part induced the reluctance to accept the final warnings when they were received. The "cry-wolf" phenomenon has rarely been more clearly demonstrated—Hitler is said to have postponed the attack on the West 29 times, often at the last minute.

Owing to their access to one of the best-placed intelligence sources of modern times, the Dutch had been correctly informed of nearly every one of these plans to attack them, from the first date selected by Hitler, 12 November 1939, to the last, 10 May 1940. Their source was Colonel Hans Oster, the Deputy Chief of German Counterintelligence, who regularly apprised the Dutch Military Attaché in Berlin of Hitler's plans—and of their postponements. Although in the end Oster provided one week's warning of the 10 May date, and there was much other evidence as well that the German attack was probably imminent, the Dutch ignored the warnings and failed even to alert their forces prior to the German attack. The Belgians, more heedful of the numerous warnings received, did place their forces on a general alert. The French, having also experienced several false alarms of a German attack, seem to have ignored the repeated warnings of their own intelligence in early May, including a firm advisory on 9 May that the attack would occur the following day. These instances also clearly demonstrate two fundamental precepts of warning made in Chapter 3 of this work: "more facts" and first-rate sources do not necessarily produce "more warning," and intelligence warnings are useless unless some action is taken on them.

The Soviet attack on Japanese forces August 1945. This is one of the lesser studied World War II examples, but clearly demonstrates the difference between strategic and tactical warning. The Japanese, who were able to follow the Soviet buildup in the Far East from December 1944 through July 1945, correctly judged that the USSR would attack the Japanese Kwantung Army in Manchuria. As noted in a preceding discussion (Chapter 16), they had also concluded by July that the Soviet troop and logistic buildup had reached the stage that the USSR would be ready to attack any time after 1 August. Despite this expectation which almost certainly must have resulted in a high degree of alert of the Japanese forces in Manchuria, the Kwantung Army had no immediate warning of the timing of the attack, which occurred about midnight on the night of 8–9 August.

The North Korean attack on South Korea, June 1950. This was a notable example of both strategic and tactical surprise, and indeed one of the few operations of this century which truly may be described as a surprise attack. Neither US intelligence, at least in its official publications, nor policy and command levels had expected the attack to occur, as a result of which there had been no military preparations for it. The South Koreans, despite many previously expressed fears of such an attack, also were not prepared and had not alerted their forces. Since strategic warning had been lacking, the short-term final preparations of the North Korean forces (insofar as they were detected) were misinterpreted as "exercises" rather than bona fide combat deployments. In considerable part, the warning failure was attributable to inadequate collection on North Korea—but the failure to have allocated more collection effort in turn was due primarily to the disbelief that the attack would occur. In addition, the "cry-wolf" phenomenon had in part inured the community—for at least a year, there had been about one report per month alleging that North Korea would attack on such-and-such a date. When another was received for June, it was given no more credence than the previous ones—nor, in view of the uncertain reliability and sourcing of all these reports, was there any reason that it should have been given greater weight. Although we can never know, most and perhaps all of these reports may have been planted by the North Korean or Soviet intelligence services in the first place. The attack is a notable example of the importance of correct prior assessments of the likelihood of attack if the short-term tactical intelligence is to be correctly interpreted.

Chinese intervention in the Korean War, October–November 1950. Among the several problems in judging Chinese intentions in the late summer and fall of 1950 was the question of the timing of their intervention. Based on the premise that the less territory one gives up to the enemy, the less one's own forces will have to recover, the Chinese can be said to have intervened much "too late" in the conflict. And this conception of the optimum time for Chinese intervention strongly influenced US judgments of their intentions. From the time the first direct political warning of the Chinese intention to intervene was issued on 3 October (to the Indian Ambassador in Peking) until the first contact with Chinese forces in Korea on 26 October, all Communist resistance in Korea was rapidly collapsing as the US/UN forces were driving toward the Yalu. As the Chinese failed to react and the Communist prospects for recouping their losses appeared increasingly unfavorable, the Washington intelligence community (and probably the Far East Command as well) became increasingly convinced that the time for effective Communist intervention had passed. In the week prior to the first contact with Chinese forces, the US national warning committee (then known as the Joint Intelligence Indications Committee, the predecessor of the Watch Committee) actually went on record as stating that there was an increasing probability that a decision *against* overt intervention had been taken.

Once the Chinese forces had actually been engaged, there was an interval of a month before they became militarily effective and launched their massive attacks in late November. Thus in this period the intelligence process again was confronted with the problem of assessing the timing of any future Chinese operations, as well of course as their scope. The four-week period produced many hard indications, both military and political, that the Chinese in fact were preparing for major military action. But there was virtually no available evidence when such action might be launched, and even those who believed that the coming offensive was a high probability were somewhat perplexed by the delay and were unable to adduce any conclusive indications of when the attack would occur. As is well known, tactical surprise was indeed achieved.

Even in retrospect, we cannot be sure whether the Chinese delayed their intervention and their subsequent offensive because of political indecision, the need for more time to complete their military preparations, or as a tactical device to entrap as many UN forces as possible near the Yalu. I believe that military rather than political factors probably delayed the initial intervention and that both preparedness and tactical considerations accounted for the delay in the offensive, but I cannot prove it. Others may argue—and they cannot be proved wrong—that the Chinese may not have decided inevitably on intervention by 3 October, and/or that negotiations with the USSR and North Korea may have delayed the intervention as much as military factors.

The Arab-Israeli Six-Day War, June 1967. There were many indications of the coming of this conflict. From 22 May, when Nasser closed the Gulf of Aqaba to Israeli shipping, tensions had been mounting, and the possibility of war was universally recognized. Both sides had mobilized and taken numerous other military preparedness measures. Before 1 June US intelligence was on record that Israel was capable and ready to launch a preemptive and successful attack with little or no warning, and that there was no indication that the UAR was planning to take the military initiative.

Inasmuch as the Israeli attacks on the morning of 5 June, and particularly the decisively effective air strikes, have often been heralded as one of the most brilliant examples of tactical surprise in this century, one may reasonably ask who was surprised, and why and in what way? The answer to the who is that the Arabs were surprised, although we were not. US intelligence predictions of the likelihood and probable success of an Israeli assault were highly accurate, although the precise timing and tactics of the operation, of course, were not known to us.

The Israelis screened their plans from the Arabs by a combination of rigid security (there was no leak of their decisions or final military preparations) and an exceptionally well-planned and effective deception campaign. There were several facets of the deception plan, one of which was to lead Egypt to believe that the attack, if it occurred, would be in the southern Sinai rather than the north. In addition, numerous measures were taken in the several days prior to the attack to create the impression that attack was not imminent. These included public statements by newly appointed Defense Minister Moshe Dayan that Israel would rely on diplomacy for the present, the issuance of leave to several thousand Israeli soldiers over the weekend of 3–4 June, public announcements that concurrent Israeli cabinet meetings were concerned only with routine matters, and so forth. In addition, the attack was planned for an hour of the morning when most Egyptian officials would be on their way to work and when the chief of the Egyptian Air Force usually took his daily morning flight.[4]

The greatest surprise in the Israeli operations was not their occurrence, however, or even their timing, but their devastating effectiveness in virtually wiping out the Egyptian Air Force on the ground. And this success in turn was due on the one hand to the excellent planning of the operation and its meticulous execution by the Israeli pilots, and on the other to the ineptitude of the Egyptian military leadership in having failed to prepare for the possibility of such a strike or to have dispersed or otherwise protected at least a portion of the air force. (It is of interest to note that the USSR, which was providing at least some intelligence assistance to Nasser, was seemingly as surprised as Egypt. One result of this was that the USSR soon began to adopt measures to reduce the vulnerability of its own air forces to surprise attack, including the widespread construction of individual hangarettes to protect aircraft.)

The invasion of Czechoslovakia, 20–21 August 1968. In the previous chapter we examined the series of Soviet decisions leading up to the invasion and the problems of determining just what decisions were taken, and when. Obviously, our perception of the USSR's decision-making process in this case has major bearing on our understanding of why the attack occurred when it did, rather than sooner or later. And, since our knowledge of the decisions of the Soviet leadership although considerable is still incomplete, we must also remain somewhat uncertain as to why the invasion occurred on 20 August rather than some time earlier that month, or alternatively why the USSR did not wait to see the outcome of the Czechoslovak party congress scheduled for early September, as many people believed that it would.

Regardless of one's views on this point, however, the invasion of Czechoslovakia illustrates some of the pitfalls of trying to assess the timing of military operations. First, we are not sure in retrospect whether the USSR was fully ready to invade on about 1 August when the deployments appeared largely completed and US intelligence concluded that Soviet forces were in a high state of readiness to invade. We do know that logistic activity continued at a high level thereafter (see Chapter 16) and that the conclusion of the so-called rear services "exercise" was not announced until 10 August. Thereafter, other military preparations were continuing, including inspections of forces in the forward area by the high command, which the meticulous Soviet military planners may well have desired to complete before any invasion. Indeed it is possible, on military evidence alone (the political evidence is less persuasive), to argue that the invasion was always scheduled by the military for 20 August, and that it was we who were wrong in our assessment that the military forces were in high readiness to go on 1 August.

It can also be argued that military factors may have prompted the invasion somewhat earlier than the political leadership might have chosen and that it was this which occasioned the leadership meetings and final decisions on 16–17 August. If so, the approaching autumn and the problem of housing Soviet forces in Czechoslovakia into the winter might have been a major factor in determining the timing of invasion.

More important, however, are the lessons to be drawn for our judgments in the future concerning the timing of operations. The Czechoslovak case well demonstrates the psychological effects on intelligence assessments when an operation does not occur as soon as we think it might, and when the community is most ready for such action. When the Soviet Union did not invade in early August but instead reached a tenuous political agreement with Czechoslovakia, a letdown occurred and intelligence assessments almost immediately began placing less stress on the Soviet capability to invade. In fact, of course, that capability was being maintained and actually was increasing. So long as this was so, the possibility was in no way reduced that the USSR sooner or later would exercise its military capability.

Above all, the Czechoslovak case provides an outstanding illustration of the critical importance for warning of the judgment of probability of attack and of the lesser likelihood that intelligence will be able to assess the timing or imminence of attack. US intelligence in this instance, as in others, placed too great weight on short-term or tactical warning, and too little on the excellent strategic intelligence which it already had. Moreover, many persons (including some at the policy level who were aggrieved that they had not been more specifically warned) tended to place the blame on the collection system which in fact had performed outstandingly in reporting a truly impressive amount of military and political evidence, much of it of high quality and validity, bearing on the Soviet intention. The intelligence community, while clearly reporting the USSR's capability to invade, deferred a judgment of whether or not it would invade in seeming expectation that some more specific or unequivocal evidence would be received if invasion was imminent. On the basis of historical precedent and the experience derived from numerous warning problems, this was a doubtful expectation; an invasion remained a grave danger, if not probable, so long as the military deployments were maintained, while the timing was far less predictable. The history of warfare, and of warning, demonstrates that tactical evidence of impending attack is dubious at best, that we cannot have confidence that we will receive such evidence, and that judgments of the probable course of enemy action must be made prior to this or it may be too late to make them at all.

North Vietnamese attacks in Laos and South Vietnam, 1969–70, 1971–72. As a final example of problems in timing, three instances of North Vietnamese attacks in Laos and South Vietnam provide quite striking evidence of the problems of assessing timing of attacks even when the preparatory steps are quite evident.

Traditionally, in the seesaw war in northern Laos, the Laotian government forces made gains in the Plaine des Jarres area during the rainy season, and the Communist forces (almost entirely North Vietnamese invaders) launched offensives during the dry season (November to May) to regain most of the lost territory and sometimes more. In the fall of 1969, evidence began to be received unusually early of North Vietnamese troop movements toward the Plaine des Jarres, including major elements of a division which had not previously been committed in the area. As a result, intelligence assessments beginning the first week of October unequivocally forecast a major Communist counteroffensive. After eight consecutive weeks of this conclusion (qualified in later weeks by the proviso "when the Communists have solved their logistic problems"), it was decided to drop it—not because it was considered wrong, but because consumers were beginning to question repeated forecasts of an enemy offensive which had not materialized yet, and the impact of the warning was beginning to fade. In mid-January, evidence began to become available that preparations for an attack were being intensified, and a forecast of an impending major offensive was renewed. The long-expected offensive finally came off in mid-February, or four months after the troop buildup and the initial prediction of the attacks. The delay was not a surprise to experienced students of the area, who had learned that the North Vietnamese meticulously plan and rehearse in detail each offensive operation and that their attacks almost always were slow in coming.

Two years later in the fall of 1971, a very similar repetition of the North Vietnamese buildup in northern Laos began, again in October and again involving the same division, although this time there were indications (such as the introduction of heavy artillery) that an even stronger military effort would be made. Intelligence assessments again forecast major North Vietnamese attacks in the Plaine des Jarres but for the most part avoided any firm judgment that they were necessarily imminent. There was almost no tactical warning of the attacks which this time were launched in mid-December in unprecedented strength and intensity. Within a few days, all Laotian government forces were driven from the Plaine, and within three weeks thereafter, the North Vietnamese launched an offensive against government bases southwest of the Plaine.

Concurrently, the North Vietnamese were preparing for their major offensive against South Vietnam which finally kicked off on 30 March 1972 after months of buildup and intelligence predictions that an offensive was coming. Initial expectations, however, had been that the attacks most likely would come some time after mid-February, possibly to coincide with President Nixon's visit to China later that month. Once again, timing proved one of the most uncertain aspects of the offensive, and we remain uncertain whether Hanoi originally intended to launch the attacks earlier and was unable to meet its schedule, or never intended the operation to come off until the end of March. In retrospect, it appears that the forecasts of another "Tet offensive" in mid-February probably were somewhat premature, since the deployments of main force units and other preparations continued through March. Nonetheless, the intelligence forecasts were essentially right, and it could have been dangerous in February to suggest that the attacks would not come off for another six weeks.

Growing Recognition That Warning Is Not a Forecast of Imminence

It is from experiences like these (which are truly representative and not selected as unusual cases) that veteran warning analysts have become extremely chary of forecasting the timing of attacks. They have learned from repeated instances, in some of which the timing of operations appeared quite a simple or obvious problem, that this was not the case. In most instances, attacks have come later and sometimes much later than one might have expected, but even this cannot be depended on—sometimes they have come sooner. But except in rare cases any forecast of the precise timing of attack carries a high probability of being wrong. There are just too many unpredictable factors—military and political—which may influence the enemy's decision on the timing and a multitude of ways in which he may deceive you when he has decided.

This experience has finally borne fruit at the national estimative level The last estimate to address possible warning of Soviet attack in Europe reversed the previous estimate (cited on the first page of this chapter) that warning of probable attack could not be given until a few hours before. It concluded instead that, once deployments and other military preparations had been largely completed, the chance of obtaining evidence of further military preparations would be greatly reduced, and that final warning that attack was imminent could likely be dependent largely on chance or other unpredictable factors.

The lesson is clear. Both the analyst and his supervisor should keep their attention focused on the key problem of whether the enemy is in fact preparing to attack at all—a judgment which they have a good and sometimes excellent chance of making with accuracy. Judgments often can be made, with less confidence in most cases, that all necessary preparations have probably been completed. A little less confidence still should be placed in forecasts as to when in the future all necessary preparations may be completed. At the bottom, and least reliable of all, will be the prediction of when the adversary may plan to strike. As a general rule, analysts will do well to avoid predictions of when precisely an attack may occur, particularly when some preparedness measures have not yet been completed. If pressed, it will normally be best to offer some time range within which the attack appears most likely, rather than attempt too specific a guess (for that is what it is). And some explanation of the uncertainties and perils of forecasting dates, backed up by historical evidence, may be helpful from time to time for the benefit of the policy maker as well.

Some official papers notwithstanding, strategic warning is not a forecast of imminent attack. *Strategic warning is a forecast of probable attack* and it is this above all which the policy official and commander need to know. If we recognize the uncertainties of timing, we will also be less likely to relax our vigilance or alerts because the enemy has not yet attacked even though he is seemingly ready.

Notes

1. Barton Whaley, *Stratagem Deception and Surprise in War* (Cambridge, Mass., MIT Center for International Studies, April 1969), pp. 177–78, and A-69. See Chapter 29 for a more detailed discussion of this work.

2. Ibid, pp. 187–188.

3. Whaley, pp. 180–181.

4. A great deal of material on the Israeli planning has been brought to light, much of it unclassified. An excellent, unclassified summary of the techniques of deception and tactical surprise has been prepared by the Syracuse University Research Corporation, Syracuse, New York.

29

Deception: Can We Cope with It?

> Stratagematic security is absolute, if the deception operation succeeds in anticipating the preconceptions of the victim and playing upon them. In that case the victim becomes the unwitting agent of his own surprise, and no amount of warning (i.e., security leaks) will suffice to reverse his fatally false expectations.[1]

CONFIDENCE THAT A STUDY of history and of techniques and principles of indications analysis will enable us to come to the right judgment of the enemy's intentions fades as one contemplates the chilling prospect of deception. There is no single facet of the warning problem so unpredictable, and yet so potentially damaging in its effect, as deception. Nor is confidence in our ability to penetrate the sophisticated deception effort in any way restored by a diligent study of examples. On the contrary, such a study will only reinforce a conclusion that the most brilliant analysis may founder in the face of deception and that the most expert and experienced among us on occasion may be as vulnerable as the novice.

The Infrequency—and Neglect—of Deception

There can be no question that deception is one of the least understood, least researched and least studied aspects of the warning problem. It has, in fact, been almost totally neglected in the training of US intelligence analysts and, even in its tactical applications, receives only scant attention in US military schools, or so I am told by those who have attended them. It is a measure of the inattention to the subject that so much of Whaley's research on the topic really broke new ground, and that some military historians have never even perceived the role which deception has played in the outcome of some major military operations.[2] When one considers the potential effects of deception on the conduct of warfare, intelligence analysis and the national decision-making process, this neglect of so important a problem becomes almost unbelievable.

One reason for the scant attention to deception almost certainly is its rarity. If true warning problems are seldom encountered, useful examples of deception are rarer still, and indeed a number of major crises of recent years seemingly have involved relatively little if any deception. A second, and related, factor is that the deception effort is likely to be the most secret and tightly held aspect of any operation and that nations often have been reluctant, even after the fact, to relax security on the deception plan, even when other aspects of the operation are fairly well known. Not surprisingly, this has been particularly true of our, security-conscious enemies. The exceptions, in which the deception operation has been recorded for our benefit and study, usually have been the result of the publication of articles or memoirs by participants in the plan, or the declassification of operational war records, usually well after the event. Thus Whaley's examples from World Wars I and II appear quite complete and reliable. In more recent cases, his unclassified data not surprisingly are quite incomplete and may even be misleading.

Deception tends to be forgotten and neglected between wars because it is not an instrument of peace. Few if any nations have made a practice of extensive or elaborate deception in time of peace, and this includes even Communist nations which we have characteristically considered to be highly devious by nature and masters of deceit and surprise. One reason that active deception is reserved for the exceptional situation—usually one in which national security interests are at stake—is that success in deception is heavily dependent on its rarity and on the prior establishment of credibility. Any nation which constantly or even frequently disseminates falsehoods would rapidly lose credibility and acceptance with other nations, and with its own populace. It is one thing to be highly security conscious and not to reveal much, and quite another to engage in an active deception effort to mislead. The most effective deceptions are by those whom we have come to trust, or at least who have been relatively truthful in their dealings with us over a period of years. Thus the true deception operation, at least a major and sophisticated one, usually is reserved only for that critical situation in the life of the nation when it is most essential to conceal one's intent. And this will usually be in preparation for or in time of war.

Deception versus Self-Deception

Still another reason for our limited understanding of deception is the interrelation of, and even confusion between, deception and self-deception. Anyone who thinks that the distinction between the two is, or ought to be, clear doesn't, as they say, understand the problem. The writer participated in a post-mortem study of the Soviet invasion of Czechoslovakia in which the analysts, drawn from the intelligence shops of several agencies, were utterly unable to agree among themselves on whether the Soviet Union had or had not engaged in deception, on whether it expected us to be deceived or not, and whether we had been the victims of self-deception. And this, of course, was in retrospect. Still less was the problem analyzed or even perceived to be a problem before the invasion. Did a majority of analysts fail to perceive the likelihood of invasion because the USSR took positive steps to deceive or mislead us, or was it only our own misconceptions of Soviet national priorities and unwillingness to accept that the Soviet Union would do such a thing, despite the evidence at hand? Space will not permit a discussion here of this fascinating and still unresolved problem. It is my own opinion that both contributed, but that we were probably more the victims of self-deception than of active Soviet deception.

The fact is that the most successful of all deception plans and operations are those which capitalize on and actively encourage the enemy to believe his own preconceptions—as Whaley notes in the quotation which opens this chapter. A similar point has also been made in an earlier section of this handbook (Chapter 10). There, it was noted that studies have shown that people do not perceive all new information objectively and that their preconceptions will sometimes lead them to ignore or reject entirely information which is inconsistent with their already formed opinions. Thus a relatively simple and unsophisticated—even obvious—deception effort may be highly effective in these circumstances in deluding the victim. And, even in retrospect, he may be unable to perceive that he was deceived by his own preconceptions as much as, or even more than, by the enemy's deception plan. With this cautionary reminder, we will proceed to examine some of the principles and techniques of active deception plans—again drawing heavily on Whaley.

Principles Techniques and Effectiveness of Deception

The principle of deception, most simply stated, is to induce the enemy to make the wrong choice; or, as General Sherman put it, the trick is to place the victim on the horns of a dilemma and then to impale him on the one of your choosing. If this is left entirely to chance, the probability of the enemy's making the right or wrong choice will be in direct ratio to the number of alternatives which he perceives as equally viable. While surprise can result from sheer misunderstanding, "the possibility of surprise through misunderstanding diminishes nearly to the vanishing point as one considers the more elaborate strategic operations."[3] Therefore, the planner must develop one or more plausible alternatives as bait for his victim and then employ a range of stratagems to mislead him. "The ultimate goal of stratagem is to make the enemy quite certain, very decisive and *wrong*."[4] If this ideal cannot be achieved (and this writer believes that it would be a rare situation in which such total deception could be achieved), the mere presenting of alternative solutions nonetheless will serve to confuse the enemy and lead him to disperse his effort or to make at least a partially wrong response:

> In other words, the best stratagem is the one that generates a set of warning signals susceptible to alternative, or better yet, optional interpretations, where the intended solution is implausible in terms of the victim's prior experience and knowledge while the false solution (or solutions) is plausible. If the victim does not suspect the possibility that deception may be operating he will inevitably be gulled. If he suspects deception, he has only four courses open to him.
>
> These are, in summary:
>
> 1. To act as if no deception is being used.
> 2. To give equal weight to all perceived solutions (in violation of the principle of economy of force).
> 3. To engage in random behavior, risking success or failure on blind guesswork.
> 4. To panic, which paradoxically may offer as good a chance of success as the "rational" course in 3.[5]

Thus, even a primitive deception effort will, by threatening various alternatives, create enough uncertainty to distract the most wily opponent and force him either to disperse his effort or gamble on being right. Further, concludes Whaley in a judgment of greatest importance for warning, even the most masterful deceivers have proved to be easy dupes for more primitive efforts. "Indeed, this is a general finding of my study—that is, the deceiver is almost always successful regardless of the sophistication of his victim in the same art. On the face of it, this seems an intolerable conclusion, one offending common sense. Yet it is the irrefutable conclusion of the historical evidence."[6]

A related, and also unexpected, finding of Whaley's study is that only a small repertoire of stratagems is necessary "to ensure surprise after surprise." The fact that the victim may be familiar with specific ruses "does not necessarily reduce much less destroy their efficacy. This can be predicted from the theory, which postulates that it is the misdirection supplied by selective planting of false signals that yields surprise and not the specific communications channels (i.e., ruses) used."[7] In other words, the same tricks can be used over and over again, and stratagem can be effective with only a small number of basic ruses or scenarios.

Whaley goes on to note that, as between security and deception, deception is by far the more effective in achieving surprise, although both may contribute and usually do. Security will also be greatly served by a deception operation since the only important security in this case will be the protection of the deception plan itself, which usually needs to be revealed only to a very small number of individuals. If the security on the deception plan is tight enough, security on the rest of the operation can be outright sloven, and "the most efficient stratagems calculatedly utilize known inefficiencies in general operational security."[8]

Whaley cites some examples of the extreme security maintained on deception plans, which the warning analyst should well heed, since it will upset all accepted theory that enemy plans may be learned from full confessions of high-ranking prisoners or defectors, or from interception of *valid* communications, authentic war plans, etc. Thus, in preparation for the Pearl Harbor attack, the Japanese Navy issued a war plan on 5 November which gave full and accurate details of the planned attacks on the Philippines and Southeast Asia but which omitted any reference to the Pearl Harbor missions of the Navy, this portion of the order having been communicated only verbally. In the Suez attack in 1956, the entire British military staff from the Allied CinC on down were not informed on the collusion of the UK and France with Israel, so tightly was this held. In the Korean war, the US planned an amphibious feint (the so-called Kojo feint) which only the most senior commanders knew to be a bluff; even the planners and commanders of the naval bombardment and carrier strike forces thought the operation was real and behaved accordingly. Thus, the misleading of one's own people has been an important feature in many deceptions, with the unwitting participants in the plan convincingly carrying out their roles in good faith and thus contributing materially to the success of the operation. So effective has security been on deception operations, that Whaley concludes that there have been almost no cases in which the deception plan itself was prematurely disclosed to the victim.

Types of Deception

This subject may be approached in a number of ways. Whaley identifies five specific varieties of military deception as follows:

Intention (i.e., whether an attack or operation will occur at all)
Time
Place
Strength
Style (i.e., the form the operation takes, weapons used, etc.)[9]

For strategic warning, the subject of this book, it will be obvious that the first of these (intention) is the most important. Indeed, some might say that this is the only variety of deception which should properly be defined as strategic, the other types above being essentially tactical problems. In fact, however, strategic warning or the perception of the enemy's intention often does fall victim to one or more of the other foregoing varieties of deception as well. Thus, in the Tet offensive of 1968, we were less the victims of misperception of the enemy's intention as such (it was obvious that attacks of some type and scope were in preparation) than of the other factors. We greatly underestimated the strength of the attacks; we were astounded at some of the places (particularly cities) in which the attacks occurred; we misperceived the style of the offensive in some degree (i.e., the extent of covert infiltration of saboteurs and troop units, again particularly into the major cities); and there was something of a misestimate of the timing of the attacks in that it was generally assumed that they would be launched before or after the holidays rather than during them (a factor which accounted for so many South Vietnamese troops being on leave and for the lax security). Thus, it was all these misperceptions of the enemy's planning and intentions which contributed to the surprise—and initial success—of the Tet offensive. We were the victims of a combination of effective security, enemy deception and self-deception.

The history of warfare is filled with examples of the achievement of surprise in time, place or strength, or a combination of them. Whaley finds that, of the examples which he studied in which surprise was achieved, the most common mode was place (72%), followed by time (66%), and strength (57%). The least frequent type of surprise which Whaley found was style, which prevailed in 25% of the cases he analyzed.[10] There are nonetheless some very famous examples, including the dropping of the first atomic weapon on Hiroshima, and the introduction of Soviet strategic missiles into Cuba.

We may close this very inadequate discussion of this approach to types of surprise and deception by observing that one of the greatest and most successful military surprises in history, the Pearl Harbor attack, involved at least four of these modes. The United States had not correctly perceived the Japanese *intention* to attack US territory at all and thus to bring the US into the war—a step which logically appeared to be a gross strategic miscalculation, as indeed it was. The *place* of attack was not perceived, since the great bulk of the evidence pointed to Japanese attacks in Southeast Asia (which were in fact initiated almost simultaneously). The *time* of the attack contributed greatly to its success, Sunday morning having been deliberately chosen because the bulk of the US warships would then normally be in port. The *strength* of the attack of course was not anticipated (since it was not expected at all where it occurred), security and deception having effectively screened the movements of the Japanese task force.

A second approach to types of surprise and deception, which is somewhat broader and perhaps more pertinent to strategic warning, is to examine the various methods or measures which may be used to achieve one or more of the foregoing types of surprise. We may identify roughly five of these:

Security
Political deception
Cover
Active military deception
Confusion and disinformation

a. *Security* in itself is not strictly speaking a type of deception, in that it involves no active measures to mislead the adversary to a false conclusion, but is designed only to conceal preparations for attack. Thus the sophisticated analyst should take care to distinguish normal or routine security measures from true deception. Nonetheless, the line between deception and security is narrow, and the two are very often confused. Moreover, an effective security program often can do much to mislead or deceive the intended victim of attack even if no more sophisticated measures are undertaken. Although security alone will not normally lead the adversary to undertake the *wrong* preparations or to misdeploy his forces, it may lead him to undertake very inadequate countermeasures or even to fail to alert his forces at all, if security is totally effective.

The Communist nations which have been our principal adversaries since World War II are, of course, proponents of the most rigid military and political security. Even the most minor and seemingly unimportant military facts are routinely considered state secrets, and all sorts of data which are revealed in the press of democratic nations are never published in Communist states—for example, the true unit designations of military units, along with their locations, strengths, equipment, and usually their commanders as well This routine security, which may seem ridiculous in peacetime, ensures that similar security will also prevail when units are mobilizing or deploying for war and makes it unnecessary to impose many new security measures which in themselves might serve to alert the adversary to an unusual situation.

Political security in closed or dictatorial societies may be even more effective. The number of persons privy to high-level decisions in Communist states is very small indeed, and we have remained in ignorance for months of major political developments or decisions in such countries. Needless to say, political decisions on the initiation of hostilities or other major military preparations are among the most rigidly restricted. We have earlier (in Chapter 10) cited the introduction of Soviet missiles into Cuba and the closure of the Berlin sector borders as two examples of the effectiveness of Soviet military security. More generally, it may be said that apparently there has never, at least since World War II, been an inadvertent leak of the specific military plans or intentions of a Communist nation. They have sometimes told us, or virtually so, what they planned to do, but there has never been a breach of security concerning such a decision, unless it was communicated to third parties. It follows that it will be most unlikely that we will be able to learn directly of such plans in the future.

Effective as such measures may be, however, there is a limit to what can be concealed by security alone, and our potential enemies know this as well.

In general, the greater the number of military measures which must be undertaken for the operation, the larger the mobilization and deployment of forces required, the less likely it is that security alone can mislead. Whaley cites the views of Clausewitz that the high visibility of large-scale operations makes their concealment unlikely, and that true surprise is therefore more likely to be achieved in the realm of tactics than in strategy. This in fact has been borne out in recent examples. Although it was possible in large measure to conceal the military deployments required for the closure of the Berlin sector borders, it was not possible to conceal those for the invasion of Czechoslovakia, and in fact the USSR made no particularly great effort to do so. Some writers have argued that modern collection systems and communications will make security measures even less effective in the future—and this would appear likely to be the case. Thus, the prospects are that various forms of active or deliberate deception will assume even more importance if surprise is to be achieved.

b. Preeminent among such methods is *political deception*—probably the easiest of all deception measures and possibly the most common. While political means may be used to promote tactical surprise, this method is of particular value as a strategic measure to conceal intent. Moreover, it is one of the most economical means of deception and one in which the likelihood of disclosure is remote, since so few people need be involved in the plan. There are a variety of political deception tactics, of which we will note a few:

The direct or indirect falsehood may be put forth through diplomatic channels, official statements, the press or other media. In its simplest and most crude form, the nation simply denies that it has any intent whatever of doing what it is preparing to do and asserts that all such charges are false—a method sometimes used, particularly if the stakes are very high. The more subtle method of the indirect falsehood is often preferred, however, and permits the leadership to maintain some degree of credibility after the event, or at least to deny charges of outright prevarication. This tactic was used by the USSR in a number of its public statements prior to the Cuban missile crisis—for example in the celebrated TASS statement of 11 September 1962 in which the USSR stated that all weapons being sent to Cuba were "designed exclusively for defensive purposes," and that there was "no need" for the USSR to deploy its missiles to any other country.

Another method of political deception which has often been used, particularly to lull suspicions in the relatively short term as final preparations for the attack are being made, is to offer to enter into "negotiations" to discuss the matter at issue when in fact there is no intention of reaching any sort of agreement. This tactic was used by the USSR on the eve of the counterattack to suppress the Hungarian revolt in November 1956, when Soviet officers opened negotiations with the Hungarians on Soviet "troop withdrawal." A form of this ruse was also used by the North Koreans for about two weeks before the attack on South Korea in June 1950 when they issued "peace proposals" calling for a single national election.

Whaley has identified a slightly different form of this deception tactic, which is to lead the enemy to believe that the firm decision to attack is actually bluff. "This is a fairly common type of ruse, one intended to restore the initiative and ensure surprise by implying that options other than war are still open, thereby concealing the full urgency of a crisis and encouraging the intended victim in the belief that he has more time and more options than is, in fact, the case."[11] He notes that this ruse was used at Port Arthur in 1904, at Pearl Harbor, in the German attack on the USSR in 1941, by the British in the attack at Alamein in 1942, and in the Israeli attack on Egypt in 1967.

A somewhat similar and relatively subtle form of political deception is to downplay the seriousness of the situation in diplomacy and in public statements in an effort to create the impression that the nation does not consider its vital interests at stake, or that its relations with the intended victim are pretty good or even improving. This may result in a quite sudden shift in propaganda to a more conciliatory tone, and friendly gestures to the adversary, after the decision or at least contingency decision to attack has already been reached. This is a quite common tactic, and one in which dictatorships, including the Soviet Union, are usually masters, particularly since their complete control of the press makes a shift in the propaganda line so easy. The

USSR employed this tactic for weeks and even months prior to its attack on Japanese forces in Manchuria in August 1945, when it undertook an ostensible easing of tensions with Japan and began to be "almost cordial" to the Japanese Ambassador in Moscow, while the buildup of forces for the attack was under way in the Far East.

The effort to deceive by political means will often entail not only the deception of many of one's own people, but may extend on occasion even to the leadership of allied nations, if the issue is of sufficient importance. And true practitioners of the art of deception even have been known to deceive their superiors (by failing to inform them of their plans)—although clearly this is a risky business undertaken only in the interests of tactical surprise for a specific military operation when war already is in progress.

c. *Cover* (here meaning the "cover plan" or "cover story") is a form of military deception which should be distinguished from active military deception, although it may often be used in conjunction with it. Cover will be used when it may be presumed that the military buildup itself cannot be concealed from the enemy, and its purpose therefore is to offer some seemingly plausible explanation (other than planned aggression) for the observable military activity. It may involve simply the putting out of false statements about the scale or purpose of the military buildup in order to conceal the real intention by attributing the military preparations to something else. Throughout history, the most usual explanation offered has been that the troops are "on maneuvers," although it is possible to think of other pretexts which might sometimes be used to explain troop movements, such as an alleged civil disturbance or disaster in a border area. The likelihood that the pretext of maneuvers would be used by the USSR to mask preparations for aggression has long been recognized by Western intelligence, and the USSR and its Warsaw Pact allies have also professed to believe that NATO exercises could serve as a cover for attack.

Despite the presumed acceptance of this principle, however, the USSR achieved at least partial success with its several announcements during July and August of 1968 that its troops were engaged in various "exercises" in the western USSR and Eastern Europe. In fact, there were no bona fide exercises and the sole activity under way was the mobilization and deployment of Soviet and Warsaw Pact forces for the invasion of Czechoslovakia. The gullibility of many analysts in swallowing, at least partially, these transparently obvious "explanations" may be attributed to various factors—among them a lack of education or experience in deception and a failure to recognize the true objective of the buildup. The USSR, in turn, had probably laid the groundwork for this gullibility by its practice, during previous years, of issuing valid public announcements concerning a series of Warsaw Pact and Soviet exercises. Thus analysts had become accustomed to accepting such announcements—which had never proved false in their experience and were conditioned to do so even when the circumstances should have alerted them to the likelihood of deception. This conditioning, interestingly enough, even extended to some studies written after the invasion which persisted in referring to the "exercises" as if they had really occurred.

d. *Active military deception* is at once the most difficult form of deception to carry out, at least on any large scale, and also one of the most effective and successful. If security and political deception measures are most effective in lulling suspicions as to intent, active military deception is the primary means whereby the adversary is led to misdeploy his forces and to prepare for an attack at the wrong place and the wrong time. Even when strategic deception has failed, or was never possible in the first place, positive military deception has proved enormously effective in achieving tactical surprise, and hence in gaining victory and/or greatly reducing the attacker's casualties in the operation. Whaley in his treatise has compiled some impressive statistics on the effectiveness and rewards of positive deception operations, some of which have been so valuable and successful as literally to affect the course of history (e.g., the Normandy invasion which is discussed in greater detail later in this chapter).

The successful military deception operation may range from a relatively simple hoax or feint to highly complex series of interrelated and mutually consistent measures all designed to create the wrong impression in the mind of the enemy (or to support his original but false conceptions) as to timing, nature, strength and place of the attack. Among the recognized techniques of active military deception are:

- Camouflage of military movements and of new military installations
- Maintenance of dummy equipment at vacated installations or in areas of the front where the attack is not to occur
- The simulation of a great deal of activity using only a few pieces of military equipment moving about
- The use of noisemakers or recordings to simulate a lot of activity
- The planting of seemingly valid, but actually false, military orders in the hands of the enemy
- The sending out of "defectors" with seemingly plausible but false stories
- The use of doubled agents for the same purpose
- The sending of invalid military messages by radio in the clear or in ciphers which the enemy is known to be reading
- The maintenance of normal garrison communications while the units themselves deploy under radio silence

- The establishment of entirely spurious radio nets to simulate the presence of forces which do not exist at all or to convey an impression of a buildup of forces in some area other than the planned attack
- A concentration of reconnaissance, bombing or artillery fire in an area other than the area of attack, or at least the equalization of such activity over a wide area so that the actual area of attack is not discernible from such preparatory measures
- False announcements or other deception as to the whereabouts of leading commanders
- Obvious training exercises for a type of attack (such as amphibious) which is not planned
- False designations for military units
- Actual deployments or feints by ground or naval units to simulate attack in the wrong area
- The use of enemy uniforms and other insignia
- Announcements that leaves are being granted on the eve of attack, or even the actual issuance of numerous passes for a day or so just prior to attack

The above list does not exhaust the tricks and ruses which have been devised and successfully used in military operations. Such active deception measures of course are often supplemented by political and propaganda deception measures, cover stories and extremely tight security on the real military operation. Thus the effect of the measures collectively can be the total misleading of the enemy as to the coming attack—even sometimes when he has accepted its likelihood and indeed may be well prepared for it in other respects. The reader is referred to some of the fascinating examples cited by Whaley and to other studies of specific deception operations.

It is obvious that a number of ruses cited above would be of limited use, and indeed could be counterproductive, in a strategic deception designed to conceal that an attack is planned at all, or in any area. In such cases, one does not wish to stir up a lot of military activity, or plant false documents about impending attacks, which will only arouse suspicions and stir the enemy's intelligence services into greater collection efforts. Some measures, such as bombing and artillery fire or even highly obvious and unusual reconnaissance, cannot be undertaken at all before hostilities have begun. For these reasons, some of the time-honored devices of military deception would not be used prior to an initial surprise attack which opens a war, the attack with which strategic warning is particularly concerned. At the same time, the reader can easily see that a substantial number of the tactics cited above could be most effectively applied to deceive us in a period prior to the initial attack.

Among the ruses which should particularly concern us are: communications deception, especially the maintenance of normal communications accompanied by radio silence on deployments; planted military orders and other documents; the use of false defectors and doubled agents; and any of the other measures which might be used effectively to distract us from concentrating on the preparations for the real attack. For we may be reasonably certain that the greater and more important the operation, the greater and more sophisticated will be the positive deception effort. The fact that we have encountered relatively few cases of active military deception since World War II should not reassure us—in fact, it only increases our vulnerability.

e. *Confusion and disinformation* probably rank second only to political deception in the ease with which they can be used to mislead and distract the opposition. Indeed, of all types of deception with which this writer has any personal experience, this tactic has proved the most effective—although this is not to say that this would necessarily be true in the future. Confusion and disinformation tactics do not have to be highly sophisticated to be successful, although of course they may be. Even an elementary program to flood the market with a mass of conflicting stories and reports can be highly effective in distracting the time and attention of analysts—and their superiors—from the reliable, hard intelligence on which they should be concentrating their efforts. Particularly if a crisis atmosphere already exists, as is highly likely, and some of the reports are sensational but have some degree of plausibility, they can prove to be a tremendous distraction. If the volume of such planted information is large enough, the analytical system can literally be overwhelmed to a degree that some important and valid facts become lost in the mill, and, others are not accorded their proper weight. There is almost no end to the damage which this type of deception can do in a crisis situation. Moreover, such a mass of material compounds immeasurably the problem of analyst fatigue, always a factor in crisis situations, and may tend to generate a series of "cry wolf" alarms which will reduce the credibility of the authentic warning when or if it is received.

The most conspicuous example in recent years of the damage that can be done by a large volume of false or unevaluated information was in the Chinese Communist intervention in Korea in October–November 1950. This is not to say that the Chinese themselves necessarily had devised a sophisticated or extensive disinformation program. It is probable that a high percentage of the mass of spurious and contradictory reports which so confused the situation and distracted the analysts that summer and fall was never planted by the Communists at all but was rather the product of the several highly productive paper mills in the Far East. The result was much the same, however, and in fact the confusion may have been compounded by our inability to tell with certainty whether a report might have been a Communist plant rather than a spurious report manufactured by some information network which had no contacts whatever in mainland China. Most of those who have examined the intelligence failure that year have given altogether too little, if indeed any, attention to the adverse effects of the volume of this spurious material on the analytical process. Regardless of the origins of the material in this case, something of the same problem could surely arise again in another crisis should our adversaries choose to exercise their full capabilities to employ such tactics.

Deception in the Normandy Invasion

"In wartime," said Winston Churchill, "truth is so precious that she should always be attended by a bodyguard of lies." The remark was made at the Teheran conference in November 1943 where the Allies began the planning for the international deception operation which was to confound the Germans in the greatest single military operation in history, the Normandy invasion (6 June 1944). Appropriately, the codename for the international deception plan was BODYGUARD. The complex series of stratagems specifically designed to mislead the Germans concerning the invasion of France carried a half dozen or more codenames. Of these the most important was FORTITUDE, the deception plan for the Normandy landings, which in turn carried the operational codename of OVERLORD.

Since the space limitations for this handbook will not permit an examination in detail of the specific tactics of various deception operations, we have selected the Normandy invasion and its accompanying and follow-on deception operations to give the reader some insight into how deception is planned and implemented and how effective it can be.[12]

It was patently impossible to conceal from the Germans that an invasion of the continent was in preparation, and more specifically to conceal the enormous buildup of troops and equipment in the United Kingdom which permitted the Nazis correctly to conclude that the main invasion thrust would be made from the British Isles. The deception operations, therefore, were designed to mislead as to the time, place and strength of the invasion, rather than the intent. The plans succeeded, almost beyond the dreams of their instigators, despite the fact that the operation involved a massive movement of ships and amphibious equipment across up to 100 miles of open water, against an enemy which fully expected the invasion, had made formidable defensive preparations for it, and had an extensive and seemingly effective espionage and intelligence service. How was this surprise achieved?

Security. Knowledge of the deception plan was restricted to as few persons as possible, with all planning papers under stringent safeguards. From mid-April onward (for approximately two months before the invasion) security measures were tightened to deny the entry of civilians into coastal areas, to prohibit military leaves outside the United Kingdom, to censor mail and news dispatches leaving the UK, and even (over diplomatic protests) to prohibit foreign diplomats from sending or receiving uncensored communications or dispatching couriers. Although there were some famous security slips, apparently the only important leak to the Germans was in Ankara, where the agent known as "Cicero" (the valet to the British Ambassador) learned the codename OVERLORD but not the date or specific area of the invasion.

Feints and Diversions. Although the Germans were unlikely to conclude that the main invasion would be made elsewhere than in France, positive deception measures were taken to encourage them to believe that secondary but major thrusts might be made against the continent in other areas—specifically in Norway and/or the Mediterranean. A variety of means was used to accomplish this, including communications deception and the planting of false information through the doubled Abwehr agents in the United Kingdom and the United States. By these means, the Germans were led to accept a false order of battle in Scotland, largely composed of what the British call "notional" (that is, non-existent) divisions, which seemingly posed an ominous and imminent threat to Norway. To divert attention to the Mediterranean, the British also carried out a celebrated ruse; they sent a British actor to Gibraltar and Algiers shortly prior to D-Day disguised as Marshal Montgomery.

Confusion and Disinformation. To confuse the enemy, the Allied intelligence services planted a vast amount of false or otherwise misleading information in German hands. Deliberate use was made of the doubled German intelligence agents, and of rumors, in order to spread all kinds of erroneous or confusing information about the date and place of the invasion. Whaley notes that, of over 250 agent reports received by German intelligence prior to D-Day concerning the invasion, only one disclosed the correct time and place—and even this one had been deliberately planted by Allied intelligence on an Abwehr agent.[13] The officer in charge of radio intercept operations for the German Fifteenth Army on the French coast had, by D-Day, become thoroughly skeptical of German intelligence: "he knew from past experience that Berlin's sources of information were inaccurate ninety percent of the time. He had a whole file of false reports to prove his point; the Allies seemed to have fed every German agent from Stockholm to Ankara with the 'exact' place and date of the invasion—and no two of the reports agreed."[14]

In this connection, mention must be made particularly of the tremendously successful British handling of the doubled German agents. The recently published Masterman and Delmer accounts have done much to explain the success of the entire deception effort. These works confirm that the British actually controlled and ran the entire German espionage system in Great Britain throughout the war; that is, the Germans received no information from their "spies" in the United Kingdom other than what the British wished them to receive. The British accomplished this extraordinary feat—which incidentally could probably only have been done in a small country lacking a land border with the enemy—by capturing apparently all of the German agents who were introduced into the country, doubling those who could be persuaded actively to cooperate or assigning radio operators of their own to maintain communications with the Germans in other cases. These operations, begun early in the war, were planned from the beginning with a view to using the agents ultimately to participate in the grand deception for the invasion of the continent, and when that time came their contribution clearly was indispensable. Masterman throughout his book constantly reminds us that the doubled agents served only as the channel to convey the false information, and that all credit for the deception itself belongs to the deception planners. Nonetheless, the foresight of the British intelligence service in perceiving the future use of these agents and their masterful handling of the individual agents must be rated one of the great intelligence accomplishments of that or any war. So successful was the British effort that those agents who did the most to put over the deception effort were, until the end, the most trusted by the Nazis.

Timing. It is not altogether clear why the Germans failed so badly to perceive the time of the invasion—particularly since there is some evidence that the time was compromised, as we shall discuss below. The most important reason, however, seems to have been that they did not believe that the Allies would invade in such poor weather—which, in fact, did postpone the invasion for 24 hours and almost caused a further delay. It was only because the consequences of a delay could be so disastrous and that the tides and moon (if it broke through) were favorable that Eisenhower reluctantly decided to go through with the invasion as planned. The Germans, however, could not conceive that the Allies would invade in such treacherous conditions. According to Ryan, "All along the chain of German command the continuing bad weather acted like a tranquilizer." German confidence was based in large part on studies of earlier Allied landing operations which showed that they had never been attempted unless the weather was very favorable. "To the methodical German mind there was no deviation from this rule . . ."[15]—an interesting manifestation of the principles of preconception and self-deception. A contributing factor also was the fact that the Germans had now been expecting an invasion for months, and as it failed to materialize, a certain degree of skepticism as to the validity of indications of its possible imminence had set in.

These factors, however, cannot fully account for the failure to have paid greater heed to the reported interception of the messages to the French underground that the invasion was imminent. For the Abwehr had obtained information in January that two lines from a Verlaine poem were to be broadcast in the clear on separate nights by the BBC, intermixed with other open code messages to the underground and a large number of meaningless messages. The first line, to be broadcast on the 1st or 15th of the month, would be a general warning that the invasion was coming soon; the second line would mean that the invasion would begin within 48 hours. The first line was picked up on the night of 1 June by monitors of the German Fifteenth Army who immediately understood its significance and promptly informed the German command in the West; this resulted in the alerting of the Fifteenth Army near the Belgian border but not of any other German forces along the coast. The second line, monitored by the Fifteenth Army a few hours before the invasion and again promptly passed on to Field Marshal von Rundstedt's headquarters, also failed to result in the alerting of German forces on the Normandy coast. As Ryan notes, it is a mystery why the intercept of this most crucial message did not lead von Rundstedt's headquarters to alert the whole invasion front from Holland to the Spanish border—the more so since the Germans claimed after the war that they had monitored and correctly interpreted no less than 15 BBC messages pertaining to D-Day.[16]

Both Allied deception and good luck apparently accounted for the extraordinary absence of so many German commanders from their headquarters at the time of the invasion. On the morning of 4 June, Marshal ("Desert Fox") Rommel, commander of Army Group B which included the forces in Normandy, left for Berlin to see Hitler and to celebrate his wife's birthday (he almost certainly had learned prior to his departure of the intercept of the first of the Verlaine lines). On the day of the invasion, all senior German commanders in Normandy had assembled, as previously planned, for war games exercises in the city of Rennes in Brittany, and they were thus some time getting back to their units. For various reasons, some other German commanders in France also were away. On the eve of the invasion, the German High Command even "decided to transfer the Luftwaffe's last remaining fighter squadrons in France far out of range of the Normandy beaches. The fliers were aghast.[17]

Deception of Place and Strength. The most elaborate, and successful, aspect of the FORTITUDE ruses was the stratagem to convince the Germans that the main invasion would be at the Pas de Calais, the narrowest point of the Channel some 150 miles to the northeast of the actual invasion sites. A largely fictitious army group was "created" for this purpose. Known as the First US Army Group (FUSAG), its headquarters was genuine but most of its formations were imaginary. In due course, the very real Lt. Gen. George S. Patton was appointed its commander. Leaks to the press and information planted through German espionage agents concerning the order of battle of FUSAG were reinforced by various positive military deception measures to create the appearance of the buildup of a massive force along the Dover coast. They included huge tent encampments and other fake installations, as well as dummy landing craft in the Thames estuary, which the Luftwaffe of course was permitted if not encouraged to photograph. This delusion was strengthened by the creation of a fictitious radio network which convincingly simulated the existence of the dummy headquarters and its units, and which, by landline routings, further created the impression that Montgomery's headquarters near Portsmouth was actually in Kent. This elaborated stratagem was reinforced by a plan of aerial bombardment which concentrated on the Pas de Calais and maintained this bombardment even on D-Day itself. On that day, an Allied invasion force of eight divisions (five moved by sea and three airlifted) was opposed in Normandy by only three under strength German divisions, so successful had the deception been. "In combination with the logical German mind, it was strikingly successful. None among Hitler's military high command in France and the Low Countries doubted that the invasion would strike the Pas de Calais."[18] Still more important, the Fuehrer himself believed it.

But this was not all.

Deception in the strength of one's forces most usually is intended to lead the enemy to underestimate the buildup, so that he may be overwhelmed with forces he did not suspect were present. In the Normandy invasion, the deception was calculated to give the opposite picture—that the Allies really had much larger forces in the UK than were actually there. And the objective of this was not only to mislead the Germans as to the place of the initial landings. The deception plan was further designed to convince the Germans that the Normandy invasion, even after it had occurred, was not the main assault but would be followed by another and still larger invasion in the Pas de Calais area using the massive and as yet uncommitted forces of FUSAG. In addition to the skillful Allied subterfuge in creating this nonexistent order of battle, at least one key German intelligence officer reportedly was so disturbed by Hitler's unwarranted optimism that he began deliberately inflating the Allied threat by uncritical acceptance of many agent reports.[19] In any event, on D-Day, when the Allies had about 39 operational divisions in Britain, the German estimate was somewhere between 70 and 93.[20] Hence, with the continuation of the Allied grand deception plan (now known as FORTITUDE II) and even though the Germans had captured two valid invasion orders on D-Day, they were successfully duped not just for days but for weeks after D-Day into believing that the main invasion was still to come in the Calais area. The hoax was successful until at least early August—primarily because the fiction of FUSAG and security on Patton's real command were effectively maintained. The result, not surprisingly, was a major misdeployment of German reserves and the immobilization in the Calais area of some 19 German divisions during this most crucial phase of the consolidation of the Allied landings in France.

The Soviet Participation in BODYGUARD in the Belorussian Offensive

The great international deception plan extended still further—to the planning for and timing of the USSR's offensive in Belorussia which was launched on 22 June 1944, or 16 days after OVERLORD. The plans for this, which had been worked out with the US and the UK, were intended to make the Germans believe that the offensive would be launched in the Ukraine rather than in Belorussia and that it would not start until July. The deception effort got under way about mid-April. Among the propaganda and public statements intended to mislead the Germans as to the nature of the summer offensive was a deceptive May Day slogan. Elaborate camouflage and security concealed the buildup in Belorussia while active deception (including dummy equipment and simulated radio networks) created the appearance of a major troop buildup in the Ukraine. Reconnaissance was carried out along the entire Western front, rather than concentrated. Only six Soviets were witting of the whole plan, and all orders were hand-delivered in a single copy.

As a result of the successful deception, the Soviet commanders in the Belorussian sector (Marshals Zhukov and Vasilevskiy) achieved a buildup of forces there one-third greater than was estimated by the Germans; it included overwhelming armored forces and permitted the Russians to obtain a superiority of up to 10-to-1 at the initial assault points. Like Rommel on D-Day, the local German commander had left for Germany for an interview with Hitler. Surprise was complete and all Belorussia was cleared in 36 days.

Soviet operations in support of the overall Allied offensive also included the attack on Finland, launched on 10 June. This was both in the nature of a feint for the later Belorussian offensive and a move to tie down German forces that might otherwise have deployed to France by misleading the Germans to believe that a major invasion of Scandinavia might be forthcoming. Of course, the attack on Finland also served other Soviet national objectives.[21]

Deception in National Doctrine and Practice

The student of warning may be only slightly encouraged, after reading the foregoing, to learn that Whaley considers that the Soviet Union is comparatively unsophisticated in the techniques and practice of deception. He attributes this in part to the stifling effect of Stalin on advances in Soviet military doctrine, and he considers that the USSR places too great weight on security and too little on active deception.

The most sophisticated practitioners of the arts of deception in World War II were the British—from whom both the US and the USSR learned some of the tricks of the trade. Not only were the British very largely responsible for such masterpieces of deception as the Normandy invasion, they also probably were responsible for much of the sophistication of the Soviet deception in the Belorussian offensive of 22 June 1944.

Whaley's treatise deals at length with German, and to a lesser extent Japanese, deception operations in World War II. Space will not permit their summary here, but any student of warning and indications will find them both fascinating and instructive.

In the period since World War II, the Israelis as well as the British emerge as the foremost and most sophisticated practitioners of deception, at least in the Western world. In this time, the Israelis have had at once the greatest need to use deception, several opportunities to practice it, and their operations (particularly in the six-day war of June 1967) have been a spectacular success. The US, which relearned the art of deception from the UK in World War II, has achieved some notable successes in deception, but, according to Whaley, lags behind both the Israelis and the British in both understanding of the art and recognition of its importance in warfare.

Based on experience with a number of operations since World War II—including several which Whaley does not treat—I am inclined to agree that the USSR and its Warsaw Pact allies have not, in these instances, displayed any great sophistication in deception, particularly in positive military deception. This could be dangerously misleading, however, since there have been few if any instances in this time which would have required an elaborate military deception effort to ensure military success—for example, it was unnecessary for either the suppression of the Hungarian revolt or the invasion of Czechoslovakia, and in the first instance there was not time to devise one. As we have noted, nations do not normally devise elaborate military deception plans except when national security interests are at stake. Moreover, the USSR has demonstrated a mastery of military security which is probably unmatched by any power in the world today—as the Cuban missile crisis should constantly remind us.

Still more pertinent may be our evidence that Soviet military doctrine places high value not only on security but also on positive deception. The USSR in 1968 released an account of the deception operation for the Belorussian offensive, discussed above, in a context which clearly indicated that its purpose was to teach the lessons of military deception to a new generation of officers. An article published in the September 1970 issue of *PVO Herald* was entitled "Ruses in Modern Battle," and it stated in part:

> In working out tasks of combat training, officers of National PVO are called upon at the same time to take into account the skillful use of ruses in modern battle.
>
> Military ruses mean the ability of the commander to hide from the enemy one's real intentions and to lead the enemy into delusion and to use this for the achievement of one's own successes. . . .
>
> In spite of the great saturation of National PVO troops with modern equipment, the role of stratagems in modern war has not decreased, on the contrary, it has grown. Stratagems help the commander hide from the enemy one's own plans and intentions, make the enemy give away his plans, tactical methods of the formation of his groupings ahead of time, and assure surprise of strike which with the dynamic, swift and decisive character of actions in modern battle plays a very important role. This is why the Minister of Defense USSR demands from the command staff that they give fundamental attention to tactical preparation, persistently teach troops to use tactical methods unexpected by the enemy, military ruses, false actions . . .

It would be foolish to assume that Soviet strategists have not studied the lessons to be learned from successful deception operations (including in recent years, those of the Israelis). We must presume that they would be capable of much more sophisticated deception than they have been willing to tell, or to show, us.

There is one type of deception which the USSR will unquestionably employ in any circumstance where it suits its interests, and that is political deception and its military hand maiden, cover stories. This is probably as near a certainty as almost any statement which can be made in the field of indications intelligence. The Soviet leadership since Lenin has shown a predilection for secrecy, security and surprise which virtually guarantees some form of political deception in any military operation. In such instances the Soviets are confirmed, if not always consummate, liars.

As for the Asian Communist nations, the combination of traditional Oriental inscrutability with Marxist doctrine and an obsession for security has for a generation left us considerably more confounded and uncertain as to their intentions, both political and military, than of those of the USSR. Indeed, in comparison, the USSR often appears downright above-board and scrutable.

Few Westerners can make any pretense of understanding the Chinese. One is reminded of the diplomat in Peking (himself an Oriental) who observed, at the peak of the Sino-Indian crisis in October 1962, that one might as well use astrology as political analysis to attempt to fathom the intentions of the Chinese. Our advantages, if any, in these circumstances may be that at least we are prepared to expect deception and deviousness from the Orientals, and that we have now had some experience in two recent wars with their tactics of military deception.

What Can We Do About It?

As should now be evident even from this brief discussion, all nations are vulnerable to deception, including even those who are sophisticated practitioners of the art themselves. Logic also suggests that, in some respects at least, democracies are likely to be more vulnerable to deception than are dictatorships and closed societies—and it is undeniably more difficult for open societies to practice it. Moreover, the history of recent years supports the judgment of experienced warning analysts that the US, at both its intelligence and policy levels, is highly vulnerable to deception.

What, if anything, can we do to make ourselves less vulnerable? Is it hopeless? The suggestions which are offered below would, I believe, be of help in real situations—although one must once again add the caution that the most sophisticated and perceptive of us also are vulnerable, and that there can be no guarantee that we will see through the enemy's deception plan. Nonetheless, despite our generally poor record, I am somewhat less pessimistic than Whaley on this score.

The first thing that is necessary, if we are to have any hope of coping with deception, is for people to learn something about it—and to study some case histories. Our neglect of this problem has been noted earlier in this chapter. It is symptomatic of this neglect that, despite over 20 years continuous experience in indications and warning, this writer knew practically nothing about the history of deception until the research for this chapter was begun. It is imperative that this subject—along with warning problems in general—be given more time and attention in both the intelligence and military schools of this nation. While we may all be vulnerable in some measure to old ruses, as Whaley maintains, we need to know what some of these ruses are if we are to have much chance of recognizing them.

Secondly, both the intelligence services and perhaps even more the policy and command levels need to understand that deception is likely to be practiced in certain situations—not only by our enemies but sometimes even by our friends. It is essential to the recognition of deception that the probability or at least the possibility of its occurrence be anticipated—or else we will almost inevitably be gullible victims of even a simple deception plan. And how can we recognize such situations? It is when great national objectives are at stake, when military forces are mobilizing and deploying, when it is clear that the adversary is "up to something." In such situations, it is the height of folly to presume that he will not also employ deception. We must be continually alert in such situations for the possibility of deception and assume its likelihood—rather than unlikelihood. Rather than wax indignant over the enemy's "perfidy," as is our usual wont, we should be indignant at ourselves for failing to perceive in advance such a possibility. Bluntly, we need to be less trusting and more suspicious and realistic.

To recognize that deception is being practiced at all may be half the battle. For the recognition of this in turn will alert us: (1) that the adversary is very likely preparing for some unpleasant surprises—else why bother to deceive us?—and (2) to start attempting to figure out what his real plans or intentions may be behind the smokescreen of the deception effort.

The easiest (or more accurately least difficult) of smokescreens to see through should usually be the political deception effort, and its accompanying military "cover story," during a period of massive military buildup. When the political conduct of the enemy is out of consonance with his military preparations, when he is talking softly but carrying a bigger and bigger stick, beware. This is the simplest and least sophisticated of deception methods. No nation should be so gullible as to fall for such tactics, without at least asking some searching questions. While the recognition that such deception is being practiced, or possibly is, will not in itself necessarily lead to a clear understanding of what the enemy is going to do, it will at least alert us that what he is going to do is probably not the same as what he says. And for strategic as opposed to tactical warning this recognition may be the most important judgment of all.

For, as Whaley notes and modern history confirms, it is virtually impossible to conceal the preparations for great military operations. The enemy, despite the most elaborate security precautions, is not going to be able to build up his forces for a major attack in total secrecy. If we are deceived or surprised in such circumstances, it will be because we either fell for his cover story or offers to enter into peaceful negotiations, allowed our preconceptions to override our analysis of the evidence, or because we were grossly misled as to the time, place or strength of the attack and thus failed to take the right military countermeasures at the right time.

As opposed to strategic warning or the recognition that the enemy is preparing to attack at all, tactical warning may be highly dependent on our ability to see through the enemy's active military deception plan. And on this score—which is largely the type of stratagem and deception which Whaley addresses in his book—experience teaches us that the chances of successful enemy deception are indeed high. Even when it is recognized that deception is being practiced—for example, if camouflaged equipment is detected this will not necessarily lead to the right conclusions as to the strength, place or date of attack. The military commander, in other words, will still have the problem of penetrating the specifics of the enemy's deception plan and preparing his defenses against it, even though the likelihood of attack itself has been generally accepted. Thus the tactical warning problem will remain even though the strategic warning problem may in large part have been resolved.

For both strategic and tactical warning, confusion and disinformation tactics present an enormous problem. The prospect that we could, in time of great national emergency, be confronted by such tactics should be a cause for grave concern. The releasing of the full disinformation capabilities of the KGB or the comparable counter-intelligence systems of other nations, together with the use of other deception techniques, could place an unprecedented requirement for sophisticated analysis and reporting on the collection mechanism, on which the substantive analyst would be heavily dependent for evaluation of the accuracy and potential motivation for deception of the informant. It is critically important in such circumstances, as we have noted elsewhere, that the collector provide as much information as possible on the origins of the report and the channels by which it was received, since the analyst who receives it will be almost completely dependent on such evaluations and comments in making his assessment. At best, it will be extremely difficult in time of emergency to distinguish even a portion of the reports which have originated with the enemy's intelligence services from those which have other origins. The tracing of the origins of rumors, for example, is often virtually impossible, yet in many cases rumors are valuable indications of authentic developments which the analyst cannot afford entirely to ignore.

In addition to the points above, there are two general guidelines which will usually assist the analyst in perceiving the enemy's most likely course of action through a fog of deception. They are:

Separate the wheat from the chaff. Weed out from the mass of incoming material all information of doubtful reliability or origin and assemble that information which is either known to be true (the "facts") or which has come from reliable sources which would have no personal axes to grind or reasons to deceive. This will allow you to establish your reliable data base which, limited though it may be, will serve as the yardstick against which the reliability or consistency of other data and sources may be judged. It sounds simple and obvious; it is usually not done.

Keep your eyes on the hardware. In the end, the enemy must launch operations with his military forces and what *they* do will be the ultimate determinant of his intent. Warning has failed in some cases primarily for lack of this concentration on the hardware. There are all kinds of ruses and red herrings, both political and military, which the enemy may devise, and they have often been highly successful in distracting attention from the all-important factor of the military capability. So long as that capability is being maintained, or is increasing, the analyst and military commander who concentrate on it are usually likely to have a much more accurate perception of the enemy's intention than are those who have permitted their judgments to waver with each new piece of propaganda or rumor of the enemy's plans.

Finally more for policy makers and commanders—the best defense of all against the enemy's deception plan may be the alerting and preparedness of one's own forces. If these are ready for the possibility of attack, no matter how unlikely that may seem, the enemy's efforts may be largely foiled even though his operation itself is really not anticipated. In other words, it is possible to be politically or psychologically surprised, and at the same time militarily prepared. The dispersal of the US fleet from Pearl Harbor as a routine readiness measure against the possibility of attack, however remote that might have appeared, would *have* saved the fleet even though all other assessments of Japanese intentions were wrong.

Notes

1. Barton Whaley, *Stratagem Deception and Surprise in War* (MIT Center for International Studies, April 1969), p. 225. This work will be published by Praeger in 1973. I wish to express my indebtedness to Mr. Whaley for many of the principles and examples cited in this chapter. This work is one of the few existing analytic studies on the nature of deception, and also contains a wealth of examples and case studies which will be a real eye-opener to any student of indications and of warfare. It is essential reading all for warning analysts.
2. Whaley claims, for example, that even analysts of that most studied of surprise attack Pearl Harbor, in nearly all cases have failed to give adequate attention to the Japanese deception effort as a contributing cause to the US failure to have recognized the likelihood of the attack.
3. Whaley, pp. 131–33.
4. Ibid, p. 135.
5. Ibid, pp. 142–43.
6. Ibid, p. 146.
7. Ibid, p. 228.
8. Ibid, p. 244.

9. Ibid, pp. 210–12.
10. Ibid, p. 214.
11. Whaley, op. cit., p. A-548.
12. When the initial draft of this chapter was prepared, information on the deception operations was available only in fragmentary and incomplete form from a number of general works on the invasion. Since that time, two books have appeared dealing exclusively with the deception plans, particularly with the British use of doubled Nazi agents to plant false information with the Germans. These two fascinating books are:

J. C. Masterman, *The Double-Cross System in the War of 1939 to 1945* (New Haven and London, Yale University Press, 1972). This is the authoritative account of the British doubling of the German agents, which was written immediately after the war and declassified by the British (with a few omissions and changes) more than 25 years later.

Sefton Delmer, *The Counterfeit Spy* (New York, Harper and Row, 1971). This work, which also appears to be reliable, covers essentially the same material as the Masterman book, concentrating on the work of the outstanding doubled agent but also providing information on other aspects of the deception operations.

Earlier works which discuss the deception operations in less detail include:

Cornelius Ryan, *The Longest Day* (New York, Simon & Schuster, 1959).

Charles B. MacDonald, *The Mighty Endeavor* (Oxford University Press, 1969)

Forrest C. Pogue, *The Supreme Command* (Washington D.C., Department of the Army, Office of the Chief of Military History, 1954).

Whaley, op cit.

13. Whaley, op. cit., p. A-373. See discussion below on timing.
14. Ryan, op cit., p. 32.
15. Ibid, pp. 79–80.
16. Ibid, pp. 31–34 and 96–97. Essentially the same account is also given by Delmer, op cit, pp. 198–99; both accounts are apparently derived from official German military records. One preeminent intelligence authority who has read this chapter, however, is skeptical that the Allies would ever have permitted the timing of the invasion to have been conveyed in advance to the French underground, given the very high risks of compromise. It is thus possible that there is some flaw in this story, perhaps even that some deception was used to lead higher German authorities to *disbelieve* the validity of the Verlaine lines. There is no indication that the Allied deception planners themselves believed that the time of the invasion was compromised. In order to maintain the credibility of their star double agent, they carefully arranged for him to convey the time of the invasion to the Germans just prior to the actual landings but too late for them to react to the news.
17. Ryan, op cit., pp. 16–21, 35–36, and 80–82.
18. MacDonald, op cit., p. 257.
19. See Delmer, op cit., pp. 144–147.

20. MacDonald, an authoritative historian for the US Army, gives 93. Whaley gives about 80, based on a variety of reports from other historians ranging from 70 to 85. According to Delmer, an official German OB estimate one month before the invasion was 75 to 80 "large formations," and one week before D-Day, Allied strength in Britain was assessed as the equivalent of 87 combat divisions. Hitler's own estimate, conveyed to the Japanese Ambassador in Berlin in late May, was that about 80 Allied divisions were in readiness to invade. It would appear that there were various German order of battle estimates—all calamitously wrong. OB analysts take note!

21. For discussion of this phase of BODYGUARD, see Whaley, op cit, pp. A-391–A-393h, and A-387–A-389.

About the Author

Cynthia Grabo received her undergraduate and graduate degrees from the University of Chicago, not far from where she was born and raised. Recruited by Army Intelligence in 1942, Grabo transferred to the Defense Intelligence Agency in 1962. She retired in 1980.

Grabo specialized in strategic warning serving as a senior researcher for the U.S. Watch Committee, and later the committee's successor organization, the Strategic Warning Staff. She is the recipient of the Defense Intelligence Agency's Exceptional Civilian Service Medal, the Central Intelligence Agency's Sherman Kent award for outstanding contribution for the literature of intelligence, and the National Intelligence Medal of Achievement. She is the author of *Anticipating Surprise: Analysis for Strategic Warning* (University Press of America, 2004), the abridged version of this book. The author resides in the Washington, DC, area.